THE UPSHIFT

Recent books by Ervin Laszlo

THE INTELLIGENCE OF THE COSMOS (2017)
RECONNECTING TO THE SOURCE (2020)
THE IMMUTABLE LAWS OF THE AKASHIC FIELD (2021)
THE WISDOM PRINCIPLES (2021)
DAWN OF AN ERA OF WELL-BEING with Frederick Tsao (2021)

Autobiography
MY JOURNEY (2021)

THE UPSHIFT

Wiser Living on Planet Earth

Handbook for urgent action

The pandemic, climate change, poverty, conflict and violence, and the refugee emergency: global crises that have an unsuspected silver lining. They bring us to a tipping point where we can choose our destiny. This book outlines the nature of that crucial point and suggests what you—as each and every conscious and responsible human being—can and must do to choose the right way forward—the way to a future of living and thriving on this small and precious but delicately balanced and vulnerable Planet Earth.

Advance Comments by International Thought Leaders

"For decades, Ervin Laszlo has taught a generation of thought leaders how to understand the problems of our contemporary civilization. In *The Upshift: Wiser Living on Planet Earth* he articulates more powerfully than ever before how to solve them. Despite our challenges there are reasons for positivity and hope, and this book provides them in abundant supply."

Marianne Williamson
Author of *A Course in Miracles*
Co-Founder, *The Peace Alliance*

"If you are looking for hope, for a "way out" of our present situation, for real solutions, and for crystal clarity in how to apply them right now, read *The Upshift* by Ervin Laszlo immediately. Your search will end—and your yearned-for future can begin—on its pages.

Thank you, Ervin, for capturing all we need to know, and sharing all we need to do, for us to produce all we need to experience in order to meet the promise of our highest and most loving-for-all tomorrow."

Neale Donald Walsch
Author of *Conversations with God*

"Your book, right now here today, is *the most* important book out there. It's a blueprint. It's not necessarily easy being the change you want to see in the world—it takes some very conscious decisions. But your book is *so* important for people today. I cannot thank you enough."

Michael Sandler
Host of *The Inspire Nation Show*

"Ervin Laszlo, our generation's greatest planetary philosopher, author of so many books that helped shape my awareness, has written a simple, clear set of steps we can all follow to create the thriving human society in harmony with our planet that, in our hearts, we know is possible."

Hazel Henderson
Author of *Creating Alternative Futures*

"Laszlo is the most prolific twenty-first century systems philosopher. His new work, *The Upshift,* may well be the most terse, elegant, and comprehensive treatise in modern times to address an ecological crisis that is, as he ably demonstrates, a crisis of human consciousness and ethics. Congratulations to Dr. Laszlo on a powerful achievement!"

Michael Charles Tobias
President, Dancing Star Foundation

"*The Upshift: Wiser Living on Planet Earth* is a perfect blend of rock-solid science, spiritual philosophy, and common sense woven into a practical guide to life in the post-pandemic world!"

Gregg Braden
Author of *The Wisdom Codes*

"A shift is imminent for our world and Ervin Laszlo has profiled this shift clearly and convincingly. *The Upshift: Wiser Living on Planet Earth* has distilled this shift into simple steps for those who choose to act. With this book, humanity has a way to respond to the call."

Frederick Tsao
Chinese Business Leader and Visionary

"In *The Upshift: Wiser Living on Planet Earth,* Ervin Laszlo urges us to seize this unprecedented opportunity to shift up to a better world. His words are compelling and uncompromising, but at the same time, filled with great love for humanity. You will hear the universe speaking through him, saying, 'Come evolve with me!'"

Masami and Hiro Saionji
Chairperson and President, The Goi Peace Foundation

"In his latest book, Ervin Laszlo offers us a thread of Ariadne, guiding us through the labyrinth of modern life to confront and overcome the Minotaur: the obsolete ideas, irrational beliefs, and outworn habits of behavior that currently control us. In the wake of our multifaceted global crisis, he shows us how to shift our thinking in practical, informed, and creative ways. A book of great importance at this crucial time of choice."

Anne Baring
Author of *The Dream of the Cosmos: A Quest for the Soul*

"We live in a time of colossal challenges necessitating systemic societal change. Dr. Laszlo's seminal new book *The Upshift* is a must-read for anyone who wishes to make a difference in our world. Drawn upon profound wisdom, it's a clear roadmap to create the much-needed shifts to build a thriving world for all."

Jean-Louis Huard
CEO of PeopleTogether.com

"As the executive producer of the documentary about the life of a genius, Ervin Laszlo, I also have the honor of calling Professor Laszlo my friend. He is a true humanitarian as well as a brilliant quantum physicist whose newest book *The Upshift* shines the light on some of the most difficult and complex issues facing the world today. The Upshift is a book that can make a real shift in the life of the reader. It is a book full of incredible wisdom and practical solutions."

Darla Boone
Executive Producer, Boone Media

"Ervin Laszlo's *The Upshift* brilliantly distils a lifetime of evolutionary thinking to address our human planetary predicament of unsustainable business as usual while proposing a detailed blueprint for a grassroots co-creative way forward to a flourishing future for life as a whole within an expanded holistic, eco-systemic, and spiritual world view—essential reading for our time!"

David Lorimer
Program Director, The Scientific and Medical Network

"Birth. Death. Rebirth. From uplift to Upshift, Ervin Laszlo is an extraordinary midwife facilitating our species' majestic rebirth with love during existential and essential times."

Alison Goldwyn
Founder & Creative Director, Synchronistory

"*The Upshift* is an inspirational rallying call and empowering guide to how we can, at this moment of our collective existential and evolutionary choice, shift both our thinking and, crucially, our behaviors. It is vital, practical and timely therapy to help lead us toward a healed, regenerated and thriving world."

Jude Currivan
author of *The Cosmic Hologram*

"Had Ervin Laszlo lived in Isaac Newton's time, the world would have faced a dilemma; whether to go the mechanistic route, or take the evolutionary path pointed towards in *The Upshift*. Not only does Ervin challenge the reductionist mindset showing the grave destruction it has caused, but he also provides an optimistic, paradigmatic direction to the future of our world. In a nutshell, 'The mystic truth' inherent in all human beings, as in all forms of life, is an active if extremely subtle "holotropism"—the tendency toward coherence and oneness," asserts Ervin Laszlo. who is among the foremost emissaries of the universe."

Sunil Malhotra
Founder & CEO, Ideafarms (India)

THE UPSHIFT

*Wiser Living on Planet Earth
Handbook for Urgent Action*

ERVIN LASZLO
author of *The Wisdom Principles*

Foreword
BY GREGG BRADEN
A Word to the Reader
BY MICHAEL BERNARD BECKWITH

Waterside Productions

Printed in the United States of America

First Printing, 2022

ISBN-13: 978-1-954968-78-3 print edition
ISBN-13: 978-1-954968-79-0 ebook edition

Waterside Productions
2055 Oxford Ave
Cardiff, CA 92007
www.waterside.com

CONTENTS

FOREWORD

BY GREGG BRADEN

In *The Upshift: Wiser Living on Planet Earth,* philosopher and system theorist Ervin Laszlo answers the question that's on everyone's mind: How do we build a better post-pandemic world while honoring the most cherished values and achievements of our past?

In eight easy-to-read chapters, this new book by Ervin Laszlo

1. identifies our once-in-a-civilizational opportunity to reimagine a healthy new world as we emerge from our global lockdown and confront the crisis of climate change;
2. zeroes in on the thinking that has led to the global crises we see in the world today; and
3. offers a blueprint for human evolution: for achieving our highest levels of individual and collective creativity.

The Upshift: Wiser Living on Planet Earth catapults you beyond casual speculation of what's possible, to reveal the practical steps that you can take right now to unveil and take in hand the collective destiny of humanity, as well as of your individual life.

A WORD TO THE READER
BY MICHAEL BERNARD BECKWITH

**Founder & Spiritual Director, Agape
International Spiritual Center**

Dear ones, you are holding in your hands a valuable and viable blueprint for humanity. Ervin Laszlo, our genius twenty-first century systems philosopher and more, has provided us with an expanded view of the unprecedented time we are living though. When you have read this book, read it again, and let it shift you into becoming the history maker the world is calling for right now. Ervin reminds us that we are to make history, not just be historians of things past. That the time we are living through now is primed for great changes—if we do not live in a state of resignation to the seeming chaos, but instead, live to catch, articulate, walk, and work toward the vision of what is possible for us all. The Upshift he is calling for is for us to see and feel what's possible, and to remember that our potential is always bigger than the problem. As I like to say, "vision without action is fantasy, and action without vision is chaos."

The Upshift: Wiser Living on Planet Earth carries great vision, but also actionable steps that we as a global community must take to leap into the next vision and version of humanity. The three clarion calls drafted by Ervin and His Holiness the Dalai Lama serve as necessary mind-shifts to embrace and manifest the next stage in our unfoldment: 1) The call for Creativity and Diversity, 2) The call for Responsibility, and 3) The call for Planetary Consciousness.

There are many systemic issues that we as a global community are facing—climate change, updating the business mindset, deepening our relationship to Mother Nature and each other. As Ervin's work shows us, they are all interconnected, creating a crisis for humanity. But as spiritual beings living this human incarnation, we must not waste this unprecedented crisis, but call for radical and immediate change!

As the old paradigms and structures are failing and falling apart, let us use this time of great flux to make the inner and outer changes that are necessary for the great leap forward! Consider that perhaps you arrived on earth in this incarnation for such a time as this to have the choice to be a part of the great Upshift. Embrace a radical ideal of what's possible and seize the moment. The time is now, and in *The Upshift* we have a beautiful, inspirational, educational, and necessary handbook.

Peace & Richest Blessings

PART ONE

AT THE TIPPING POINT

Chapter 1
Now we can Shift to a New World

The Opportunity of the Global Crisis

The Old Testament told us, where there is no vision, the people perish (*Proverbs* 29:18). To live and live well, we need vision. This requirement is sadly lacking in the contemporary world. Helplessness, dejection, and disorientation abound. The vision most people have of the future is of crisis and breakdown. The clouds on the horizon obscure the sun that still shines above in the sky. The sun does shine, but we have to perceive it to be guided by it. It's high time to look at the bright alternative and not just at the dangers.

The chances of shifting to a new world have never been as good as they are today. We are at the tipping point. The old world is moving into history, and a new world is waiting to be born. It will be a different world, but whether it will be better or worse than the world we have today is yet to be decided. In the final count, the decision rests with us—with you and with me, and with all awakened and responsible people in the human family.

Where We Are Now: A Balance Sheet

Shifting UP to a better world—or drifting DOWN to crisis and chaos: this, as Hamlet would say, is the question. It is the challenge of the tipping point at which we find ourselves. The crises we have been experiencing are driving us to change, and it is no longer a question of whether to change or not to change, only *how* to change. The wrong way would have catastrophic consequences. We would encounter a series of deepening crises—the virus crisis and the climate crisis would be only forerunners. But the right way could open an era of responsible living on the planet. We need to face the crucial questions: *Where are we today, and what is this tipping point at which we find ourselves?*

There are signs that we may be pulling out of the crises on the right side of the tipping point. The physical conditions for human life on the planet are becoming known. They are not only known, but are actually acted upon. The actions contemplated today may or may not be sufficient to shift to a better world—but there is a reasonable chance that they may be.

There are positive developments in the human world, as well as persisting problems. Let us draw up a balance sheet. On the positive side, we can enter that there are major projects to ensure the conditions under which humanity can persist on earth. The international community is waking up. An unprecedented set of remedial measures are being developed, and major pledges are made to meet the global crisis. Attention is focusing on the United Nation's Sustainable Development Goals, on the climate objectives brought forth by the international climate conferences including COP 26, and on the Resolutions of the G20, the club of industrialized nations. It is becoming clear that reaching these and related objectives is essential for ensuring the persistence of now nearly eight billion humans on the planet.

On the negative side, we must note that the financial costs of achieving the goals and objectives are not assured. It is a good thing to have pledges, but we must also come up with the financing to achieve them. So far, neither the international community nor the business world have assured adequate financing. At the same time the threats keep intensifying, and the costs of facing them keep rising.

On the negative column of the ledger, we should add that even if the pledges were fully financed, the goals to which they are dedicated would not entirely overcome the threats. More is needed to prevent a catastrophic rise of the level of the world's oceans and the aridity of vast tracts of the continents than creating carbon neutral economies by 2050, not to mention only by 2060, and thereby stop global warming at the level of 1.5 degrees. While achieving this goal is basic, achieving it does not foreclose the further negative manifestations of climate change, but would merely mitigate them and postpone their coming.

Where Are We Going?

This is where we are now. We have agreement on a number of crucial goals and objectives, and partial financing to achieve them. We are moving in the right direction, but are not moving far enough nor rapidly enough. To meet the challenges of global sustainability, bolder and more adequately financed measures are needed. Resolutions and projections on national and international levels are no longer sufficient. Action is needed on the global scale, and this presupposes the agreement and active participation of a critical mass of the world's peoples.

The conscious involvement of the people of the world remains a basic precondition of heading off the advent of unsustainable and ultimately unlivable conditions on the planet. The people factor remains crucial. Notwithstanding the recurrence of dictatorships and the persistently hierarchical way in which power is applied in nationalistic societies, the overall system of humanity now has a democratic bottom line. In the final count, in most countries it is the people who decide. They have a word in electing their leaders, and they can choose the services and products placed on the market by business enterprises.

The question is, what is being done to ensure the active engagement and support of a critical segment of today's people for projects to create sustainable conditions on the planet? For the present, enlisting the engagement of people at the grassroots level commands the attention only of concerned scientists and humanists and some social media, and not of mainstream media, global business enterprise, and national politics. For lack of global-scale public engagement, the sword of Damocles remains suspended above our heads.

Where we are going will depend on our recognition that we need a new mindset among the people. Local and global unsustainability are the result of a flaw in the dominant mindset. Building an effective political, social, and ecological system to replace the fragmented malfunctioning system now in place calls for widespread popular participation. This presupposes a mindset revolution among the people of the world. The timely advent of this revolution will enable the human community to find the right way beyond today's crucial tipping point.

Three messages

Here are three messages that highlight the issues with uncompromising clarity. One warns us of the dangers that face us, the second calls for an almost messianic revolution, and the third points to our powers to overcome the dangers.

> *In the face of an absolutely unprecedented emergency, society has no choice but to take dramatic action to avert a collapse of civilization. Either we will change our ways and build an entirely new kind of global society, or they will be changed for us.*
>
> Gro Harlem Brundtland, former
> prime minister of Norway,
> Chair of the World Commission on
> Environment and Development

> *Ultimately, this Earth can be saved from mankind only if people are prepared to live with nature, rather than upon nature. Recognition of the oneness of life on earth, of its beauty and its sanctity, must be spread by an almost messianic revolution.*
>
> Sir Mark Oliphant
> former governor of South Australia,
> founding member of the Pugwash Club*

---

* Sir Oliphant's message was transmitted in a letter to Michael Ellis of Australia. I am grateful to Dr. Ellis for sharing it with me.

Here and there in history one notes a sudden concentration of energies, a more favorable constellation of social opportunities, an almost worldwide upsurge of prophetic anticipation, disclosing new possibilities for the race: so it was with the worldwide changes in the sixth century BC symbolized by Buddha, Solon, Zoroaster, Confucius, and their immediate successors, changes that gave common values and purposes to people too far separated physically for even Alexander the Great to unite them. Out of still deeper pressures, anxieties, insecurities, a corresponding renewal on an even wider scale now seems open for mankind.

—Lewis Mumford, Historian, in his
classic *The Conduct of Life*

Yes, Madam Brundtland, we are living an unprecedented emergency. Yes, Sir Oliphant, in today's world a revolution must spread that is almost messianic. And yes, Professor Mumford, we do possess precious, and in the contemporary world still largely ignored, capacities for overcoming the emergency. There is a force in the universe that is with us.

May the force be with you!
Star Wars greeting

Shifting Up: The Force Is with Us

The force that empowers us to shift beyond the challenges and crises of our time is not science fiction, nor is it one of the forces of classical science. It is a subtle inclination, an almost spiritual leaning and tendency. In the language of the life sciences, it can be called "a tropism," an attraction toward particular resources or conditions. In the context we adopt here, which is that of the systems sciences, it is an "attractor." Attractors are conceptual tools for measuring the state of a complex system, assessing the configuration of the forces that maintain it and evolve it.

The attractor active in nature is a subtle bias for particular outcomes among otherwise nondirectional or possibly random interactions. This bias is toward the formation of integral ensembles of diverse elements—complex and coherent entities. They possess a degree and form of integration that amounts to "wholeness." The attractor that biases interactions in the universe is wholeness-oriented: it is a "holotropic attractor." It need not be a separate, additional law of nature: it could be, and most plausibly is, the emergent effect of the known laws.

The holotropic attractor appears as a tropism toward coherence and oneness throughout nature. In the human realm, it is present on the level of instinct and intuition, or subtle spiritual insight. It connects the individual with nature and the universe. It is effective whether we are aware of it or not. But it is more effective when we are aware of it because then we can consciously align with it.

The existence of an evolutionary bias or tropism in nature is an ancient insight, it is not discovered, but rediscovered today. The implications of this rediscovery are profound. They suggest that the evolution of life is not a response to external conditions, but the unfolding of an attraction intrinsic to the universe. There is evidence to support this rediscovery.

Now we know that for nearly two hundred thousand years, the human genome has been essentially the same as it is today. The genes responsible for the functioning of our organism, coding such advanced faculties as articulated speech and a self-reflective mind, are being identified. These genes have been part of our genetic makeup for two hundred thousand years or more. They were present in our genome *before* the faculties they code would have been expressed in our phenome (our biological organism). How come they were part of our genome? Their presence could not have been due to the interaction of many generations with the world around them: they *predate* such interactions. Their presence is also unlikely to be due to a serendipitous series of chance interactions, as the probability of their being constituted by random interactions is astronomically negative. Not just two hundred thousand years, but the 3.5 to 4 billion years that elapsed since the beginning of life on the planet would not be sufficient to provide a reasonable probability that our genes would have been created by a random rearrangement of their cells and molecules.

The conclusion is inescapable. There is a force in the universe— a tropism or attractor—that biases no-longer random interactions toward forming complex entities—entities that are coherent enough to maintain themselves and to evolve toward higher levels of coherence and complexity. This force must have been present in the universe at the time the first ensembles of quanta formed in the primeval "soup" of the early universe. Quantum physics discloses that neutral atoms, the structural basis for molecular and macromolecular ensembles, did not arise from random processes: the universe has been "programmed" from its very beginnings to produce complex and coherent entities. These entities emerge whenever and wherever conditions are suitable to form the complex chains of carbon on which life is based. As the surprisingly widespread presence of the "extremophiles" testifies, life emerges even in such unlikely places as the bottom of the sea, near active volcanos, and in the vicinity of active stars. It emerges under conditions of high salinity, extreme aridity, and nearly lethal levels of radiation.

The insight we now come to is that evolution in the universe is neither random, nor externally generated. It is catalyzed by a cosmic evolutionary force: a tropism or attractor present in space and time, and hence present in you and me. This is "The Force" young people wish to be with them. The cosmic attractor is that force, and it is with them, the same as it is with all living things in the universe.

Science fiction, as well as ancient wisdom, are often the precursors of legitimate science, and legitimate science often rediscovers creative fiction and ancient wisdom. This is the case in regard to the tropism inherent in the universe. The existence of this force is re-stated today in imaginative fiction, and it was recognized in the spiritual traditions for thousands of years. It was articulated by Lao Tzu in the first century AD. The twenty-first verse of the *Tao Te Ching* states that,

The Great Virtue is to follow the Tao and only the Tao. The Tao is shadowy and intangible.
Intangible and evasive, and yet within it is a form. Evasive and intangible, and yet within it is a substance. Shadowy and dark, and yet within it is a vital force.
This vital force is real and can be relied upon.

In the thirteenth century, the Japanese Buddha Nichiren Daishonin called the vital force inherent in things "the mystic truth." In his *On Attaining Buddhahood in This Lifetime,* he wrote,

If you wish to free yourself from the sufferings of birth and death you have endured since time without beginning and to attain without fail unsurpassed enlightenment in this lifetime, you must perceive the mystic truth that is originally inherent in all living beings.

The rediscovery of the evolutionary force in nature is essential wisdom at today's tipping point. It empowers us to upshift to a better world—when we recognize it, and align with it.

ANNEX: An Excursion into Contemporary Systems Theory

Today, we are at a tipping point. If we are to upshift to a better world, we need know something about the what is required to pull out of the tipping point on the right side and create a better world. This excursion into modern systems theory is offered to provide an answer. It outlines the principal elements of the practical wisdom we need.

Our world is in crisis, and crisis in a system can be described as a "bifurcation"—a forking in the trajectory of the system's evolution. This can be a good thing: it can be the passport to the system's continued evolution. Or it can be a bad thing, an overture to the system's demise.

When a system becomes unstable, it is compelled to shift the trajectory of its evolution. A human system in the real world either bifurcates, or leaves the scene of history. The requirement is fundamental change, and fundamental change calls for crisis—it does not occur in a functional system. Crisis is a disruption in the functions of the system. The disruption, and the crisis it creates, are critical fluctuations in the existence of the system. When the fluctuations reach the tipping point, the system either shifts its evolutionary trajectory to a more viable branch, or it decomposes to its individually stable components.

The system in question may be a group of cells in an organism, a species in the biosphere, a society of human beings in the global community, or even a solar system in the galaxy. The system that now concerns us is the system formed by humanity: the ensemble of the social, economic, and political systems we have created on the planet. This system manifests critical fluctuations and is increasingly unstable. Its evolutionary trajectory is on the point of bifurcating.

Insightful people have been anticipating that the human system on the planet will bifurcate—sooner or later, its trajectory will break down. Systems scientists have been investigating the dynamics of this process and reached some important conclusions. It turns out that the process of bifurcation is one-way; it cannot be reversed. But it is not predetermined—it allows for choice. Even a small fluctuation within the system can "nucleate" and govern the evolution of the whole system. In today's world there are many fluctuations—tipping points—that decide the choice among various alternatives. Bifurcation in a complex system is governed by "chaotic attractors"— a probabilistic process of change across chaos. This means extreme sensitivity to change, and significant openness in the system to alternative paths of evolution.

Living at the tipping point of a bifurcation, we are granted an opportunity. Given sufficient instability in our political, social, and ecological systems, even a small group of dedicated people can influence—even critically influence—the way our system evolves. Even a small group of people can create the small but potent "initial kick" that chooses among the alternative evolutions opened by the bifurcation and launches the system on a more stable and promising trajectory.

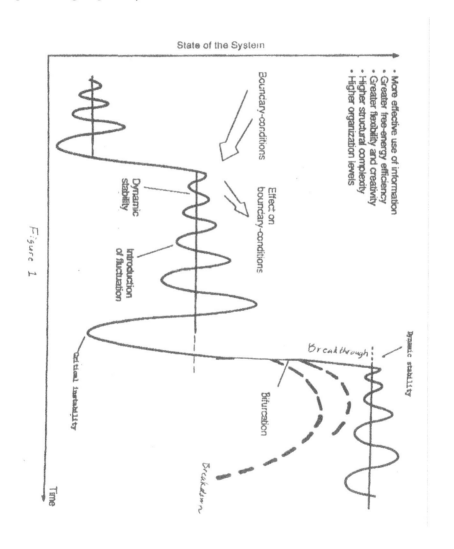

Figure 1

Legend to Figure 1

A typical sequence of events in a process of bifurcation. (1) Fluctuations augment in the system. (2) The fluctuations reach a critical threshold where the system no longer returns to its previous state but leaps to a new state of dynamic stability—or breaks down. (3) In interaction with its environment, a series of intensifying fluctuations appear in the system. (4) A fluctuation reaches a further critical threshold—the "tipping point"—and shifts the system to a new state of dynamic stability.

The iteration of the process of bifurcation either leads to the system's demise, or shifts it to higher states and forms of dynamic stability. The resulting system is characterized by a more effective use of information; greater efficiency in the employment of the available resources; greater complexity in structure; and further increase in its levels of organization.

The difference between bifurcation in a system composed of human beings and bifurcation in every other system is the presence of an evolved consciousness. In the human system, the participants can be aware of what is happening and can consciously act to influence it. The decisive actions do not necessarily call for the exercise of brute force; they can be created by a shift in the mindset of people. Victor Hugo said that there is nothing as powerful in the world as an idea whose time has come. A shift in the mindset of a critical mass can be powerful enough to decide the way the human system evolves—whether it shifts down to chaos and decay, or up to a more stable condition.

In the absence of the positive "kick" administered by people with a new mindset, the chances that the human system would evolve toward a more table and humane condition are slim. An initial kick in a destabilized system is the impetus needed to overcome resistance to change in the system. As long as the system appears salvageable, those who have a significant stake in it are reluctant to abandon it in favor of something new—they fear for their privileges.

They would rather try salvaging the existing system than reaching out to an alternative.

Resistance to change cannot be ignored: it blocks the evolution of the system. But the source of the resistance and be identified, and resistance can be overcome, or at least mitigated by purposive action. This is all the more possible because fundamental change seldom comes from the top. It comes from the bottom. The top becomes destabilized and loses its power to govern the system. At first marginal, but then more and more potent changes at the bottom take over the governing of the system. This is the way evolution unfolds in nature, and it is likely to be the way it will take place in the human world as well.

We are at a tipping point; we cannot stay as we are, and we cannot go back to how we were. But we can go forward. Given the opportunity offered by the bifurcation, we can create the "kick" that launches the human community on a positive path of its evolution.

CHAPTER 2

NO WAY BACK

Let's stop for a moment before plunging into the enterprise of upshifting to a better world. Let's first take a cool look at the world we did in fact build. What kind of a world is it? What lessons does it hold for an upshift to a better world?

❊   ❊   ❊

Lessons of the Fateful Business of Business as Usual

Critical Conditions in the Global Ecology

In the late twentieth and early twenty-first centuries, conditions of a less and less tractable kind have surfaced in relations between people and nature—between the human system and the planetary ecology. The advent of these critical conditions has not been accidental; it could have been foreseen, but alas it seldom was. The key to understanding the advent of these conditions is tracing them to two long-term trends:

1. the rapid growth of demand for the planet's physical and biological resources by a growing human population, and
2. the concurrent shrinking and threatening depletion of some of the planet's essential physical and biological resources.

If these trends were to continue to unfold, the curves of their unfolding would cross, and humanity's demand on the resources of the planet would exceed the planet's capacity to meet it. This would create a historically unprecedented condition.

For most of our five-million-year-plus history, humanity's demand in relation to the available resources has not been significant. With primitive technologies and smaller numbers of people, planetary resources seemed limitless. Even when we exhausted a local environment and depleted local resources, there were always other resources and environments to exploit.

But by the middle of the nineteenth century our population reached one billion, and it is over 7.8 billion today. The world population may grow to ten billion by the middle of this century. Approximately 95 percent of this growth would occur in the presently poor countries and regions, but these regions are not a closed system: their penuries would soon be distributed to all the countries and regions of the globe.

Today's humans constitute only about 0.014 percent of the biomass of life on earth, and 0.44 percent of the biomass of animals. Such a small fragment need not constitute a threat to the entire system, and hence itself. But because of excessive resource use and environmental degradation, we do threaten the global system. Our impact on the earth's resources is entirely out of proportion to our size. We cannot continue to increase our demands without provoking a major dieback of our populations.

One of the most important measures of the human impact on the planet, the "ecological footprint," gives us some estimates. The ecological footprint measures the share of the planet's biological productivity used by an individual, a city, a nation, or all of humanity. It is the area of land required to support a human community. If the footprint of a settlement is larger than the area of that settlement, that settlement is not independently sustainable. A city is intrinsically unsustainable because very few of the natural resources used by its inhabitants come from within its boundaries—most of them (such as food, water, and waste disposal) rely on hinterlands and catchments. But entire regions and countries could well be sustainable: their ecological footprint need not extend beyond their territories. This, however, is not the case. Today, humanity lives beyond its means—it exceeds the capacity of the planet to maintain a supply of breathable air and drinkable water, absorb pollution, and provide room for human habitations and their essential infrastructure.

The situation would be still more dramatic if all countries would adopt socioeconomic development on the Western mold. If the footprint of the more than forty "rich" countries was to be attained by all 189 formally constituted nation-states of the world, the global overshoot would be at least 100 percent. This means that if humanity were to remain in balance with – its ecological base, we would have to colonize another planet with endowments similar to Earth.

Our ecological unsustainability is the result of a mode of development as old as human civilization. Prehistoric societies were stable and enduring; they evolved a sustainable relationship with their environment. Only the energy of the sun entered the nature–human system, and only the heat radiated into space left it—everything else was cycled and recycled within it. Food and water came from the local environment and were returned into that environment. Even in death the human body did not leave the ecology; it entered the soil and contributed to its fertility. Nothing that men and women brought into being accumulated as "nonbiodegradable" toxins; nothing we did caused lasting damage to nature's cycles of generation and regeneration.

The situation changed when groups of early humans learned to manipulate the environment and broke open the loop of regeneration that earlier tribes maintained. With this change, the human impact on the natural environment began its fateful increase.

As better tools were invented, more resources could be accessed and existing resources could be further exploited. As a result, more people could be supported, and the number of humans on the planet could grow. With the control of fire, perishable foods could be maintained over longer periods, and human settlements spread over the continents.

No longer content to gather and hunt their food, humans learned to plant seeds and use rivers for irrigation and the removal of wastes. They domesticated some species of dogs, horses, and cattle. These practices further increased humanity's impact on nature. Nourishment began to flow from a purposively modified environment, and the growing wastes from larger and technologically more advanced communities continued to disappear conveniently, with smoke vanishing into thin air and solid waste washing downstream in rivers and dispersing in the seas. If a local environment gradually became arid and inhospitable—due to deforestation and overworking the soil—there was always virgin land to conquer and to exploit.

This is no longer the case. Now we are destroying even islands, although it is evident that from them we cannot move further: they are finite both in space and in resources. We have been overexploiting the resources of islands for centuries with devastating consequences—Easter Island is a well-known case in point. Now powerful technologies are employed to exploit island resources, and here the island of Nauru is a striking illustration. An island republic of fewer than eleven thousand people, Nauru was a tropical paradise but a few decades ago. Then international mining companies began to extract phosphate, stripping the topsoil. As the price of phosphate rose from ten dollars a ton to over sixty-five dollars, by 1968 the island republic became the second-richest country on earth. But the ecological price has been high: in the span of a few decades, the island became a moonscape of gnarled spiky rock. Its economy has been failing. Its inhabitants turned from eating vegetables and other fresh produce to fatty and salty tinned goods, producing one of the highest levels of obesity, heart disease, and diabetes in the world.

Now the island republic seeks further income from opening up the seabed for exploration—an economically promising scheme as the resources to be won from the sea bottom are becoming more and more valuable, needed for the production of a new generation of batteries and other advanced technologies.

Today even the deep sea faces the prospect of large-scale human intervention. The results, experts admit, are hardly foreseeable.

Due to unreflective and irresponsible interventions in the delicately balanced ecology of the planet, we are now approaching the outer limits of the earth's capacity to sustain human life. This capacity is becoming more and more reduced due to irresponsible interventions. This fact is obscured—and purposively disregarded—by the short-term economic gains offered by the interventions. The price is clearly evident in the field of agriculture. Chemically bolstered mechanized agriculture is a lucrative enterprise, it increases yields per acre and makes more acres available for cultivation, but it also increases the growth of algae that chokes lakes and waterways. Chemicals such as DDT are effective insecticides, but they poison entire animal, bird, and insect populations.

Waste disposal contributes to the reduction of the planet's human carrying capacity. We discard not only our biological wastes into the environment. We inject over one hundred thousand chemical compounds into the land, rivers, and seas; dump millions of tons of sludge and solid waste into the oceans; release billions of tons of CO_2 into the air; and increase the level of radioactivity in water, land, and air.

The wastes we discard into the environment do not vanish; they plague people near and far, including those who produce them. Refuse dumped into the sea returns to poison marine life and infest entire coastal regions. In the industrialized countries, over a million chemicals are bubbling through the groundwater systems; and in many countries, even in the least developed ones, rivers and lakes have up to a hundred times the acceptable level of pollutants. The smoke rising from homesteads and factories does not dissolve and disappear—the released CO_2 remains in the atmosphere and affects weather the world over. Not surprisingly, there has been a massive increase in allergies in both urban and rural populations. The appellations of toxic environmental effects constitute a whole new vocabulary: there is MCS (multiple chemical sensitivity), wood preservative syndrome, solvent intolerance, chemically associated immune dysfunction, clinical ecology syndrome, chronic fatigue syndrome, fibromyalgia, sick-building syndrome, and many more.

Critical Conditions in the Social and Economic Domains of our World

The advent of critical conditions in relations between the system of humanity and the global ecology has been accompanied by the advent of critical conditions in the human system itself. The human population has been exposed to growing stress. While economic, ecologic, and technological globalization proceeded at a breakneck pace, a third of the economies of the world have been left out of it. The planet-wide network of states has been growing together in some respects—first and foremost in trade and communication—but it has been coming apart in the social and economic domains.

As already noted, pursuing economic growth without checks and balances is a dangerous option. Economic growth has been made possible by the spread of information and communication technologies, and the dynamism of the private sector. But this high-impact process did not produce well distributed benefits. Many states and economies have been left out of it, and their populations have been resentful, often turning violent.

The underprivileged segments of the population have been facing less and less tenable conditions. The nodes of poverty have been expanding, increasing the gap between rich and poor. The so-called Gini coefficient, a measure of the rich–poor gap, has been growing. The poor countries and poor populations have been hit hardest by the crisis and the recession it created. Vast populations have been pushed to the very limits of physical subsistence.

The gap between rich and poor has also been growing *within* given states and regions. The poor segments of most countries have been growing, and their poverty has increased. This is a major injustice, and the poor countries and populations are deeply resenting it, sparking an ever more violent protest movement.

There is good reason to protest. If access to the planet's physical and biological resources were evenly distributed, all people and populations could be supported. If food supplies, for example, were equally accessible, every person would receive about a hundred calories more than it is required to replace the 1,800 to 3,000 calories each normal person expends every day (the average healthy diet is about 2,600 calories). But people in the rich countries of North America, Western Europe, and Japan obtain 140 percent of the caloric requirements of a normal healthy life, whereas people in the poorest countries, such as Madagascar, Guyana, and Laos, are limited to only 70 percent.

The world's pattern of energy consumption is just as disparate. The averages tell the story. The average energy consumption in Africa has been half a kilowatt hour of commercial electrical energy, while the corresponding average in Asia and Latin America has been 2–3 kWh. The average in North America, Europe and Australia rose to 8 kWh. With 4.1 percent of the world population, the United States alone has been consuming 25 percent of the world's commercial energy.

Affluent and wasteful consumption in the rich parts of the world is not the only cause of the crisis of the international system; the way poor people often obtain the resources required for their survival is unsustainable in itself. The more than a billion and a half people who, according to World Bank estimates, live at or below the absolute poverty line (defined as the equivalent of one dollar a day) destroy the environment on which they vitally depend. This creates a major demographic unbalance. With rural environments degrading, people abandon their native towns and villages and flee to the cities.

Urban complexes have experienced explosive growth—one out of fewer than three people now lives in a city, and by the middle of the century, two out of every three people will do so. By then there would be more than five hundred cities with populations of over one million, and thirty megacities exceeding eight million. Such cities cannot be self-sustaining; their ecological footprint vastly exceeds their territory. The bigger they are, the greater their dependence on their hinterlands.

Social and cultural disparities and conflicts have been stressing life in most societies. Traditional social structures have been breaking down; the family, as sociologists say, has become "defunctionalized." The traditional functions of family life have been taken over by institutions dominated by outside interests. Child-rearing has been increasingly entrusted to kindergartens and company or community daycare centers. Leisure-time activities have been dominated by the marketing and PR campaigns of commercial enterprises, and the provision of daily nourishment has shifted from the family kitchen to supermarkets, prepared food industries, and fast-food chains.

In cities, the exigencies of economic survival and the striving for a higher standard of living have broken apart the traditional extended family, and in regions of extreme poverty, even the nuclear family has proven difficult to keep together. To make ends meet, women and children have had to leave the homestead and search for ways to generate money. Women have been extensively exploited, given menial jobs for low pay, and young people have fared even worse. More than fifty million children have been forced into manual labor, often for a pittance, working in factories, mines, and on the land. Many are forced to live on the streets as "self-employed vendors," a euphemism for beggars.

An even more deplorable consequence of family poverty has been the letting go, and sometimes the outright selling, of children into prostitution. UNICEF called this "one of the most abusive, exploitative, and hazardous forms of child labor." In Asia alone, one million children are believed to work as juvenile prostitutes, exploited by the highly profitable and growing industries of international pedophilia, fueled by widespread sex tourism.

Whether in the cities or in the countryside, poverty has been characterized by malnutrition, joblessness, and unjust and degrading conditions of life. At the same time it has made for the over-working of productive lands, the contamination of rivers and lakes, and the lowering of water tables. This has created a vicious cycle. Poverty prompted poor families to have many children because children can help families at the level of subsistence to garner the resources needed for their survival. The resulting growth of poor populations has both inflicted further damage on the environment, and destroyed the kinship structures on which the stability of traditional societies has always depended.

The Bottom Line

Business as usual created a world that is ecologically unsustainable and socially and economically unjust. The poor populations are locked into a vicious cycle of poverty creating depressed conditions, and depressed conditions expanding the nodes of poverty. In turn, the more affluent segments find themselves living and working under conditions of no-holds-barred competition and job uncertainty, producing more and more anxiety and stress. The specter of deepening crises appears ever more insistently on the horizon.

The business of business-as-usual proved to be a fatal business; it created an unsustainable, unjust, and crisis-prone world. Now, when the global crisis grants us the opportunity to build a different world, it's time to ask ourselves: Is the world of business-as-usual the world we want to keep? Is it the world we want to hand down to our children?

This may be a moot question in the end. We are living in the midst of the sixth mass extinction, known as the Anthropocene extinction, because it is triggered by humans. Since the middle of the twentieth century, more than 60 percent of the animal species on the planet have disappeared, and according to some ecologists the remaining 40 percent may follow in not much more than twenty years. That could lead to the destabilization and eventual collapse of the global ecosystem. The dinosaurs disappeared after millions of years of reign following the collapse of the global ecosystem (probably due to the impact of a giant meteorite) and our arbitrary interventions in the planet's ecology could have similar consequences for homo sapiens—who would then prove to be not sapient at all.

But we are not extinct yet, and we could change direction. There is no way back, and no future in continuing to go the way we have been going—but we could find the way to move forward. This is the option we shall explore in this book.

CHAPTER 3
THE WAY FORWARD

We have lived unsustainably and irresponsibly on this planet, damaging the integrity of both nature and society. The time has come to adopt more responsible ways of living. This calls for fundamental change—for fundamentally new ways of thinking and acting.

In the past, new ways of living evolved over many generations. The rhythm of change was relatively slow and allowed people to adapt their relations to each other and their attitudes toward their environment. These times have passed. The crucial period for finding our way to a better world is now compressed not just into the lifetime of this generation, but into the span of a few years.

We could afford the luxury of moving forward through trial and error in the past, but we cannot afford to do so today. We need to go where we want to go, and how we can go. Climate change and the virus pandemic came on top of a number of critical conditions in the social, economic, and ecological spheres. If not checked, these conditions would give rise to equally if not much more serious conditions—to ever deepening local and global crises.

The overexploitation of resources and the impairment of nature, coupled with the unequal distribution of wealth and power, call for fundamental change. The way forward is not a return to the way already traveled—to the world of business-as-usual. There must be a better way to take us forward.

The better way is very different from the way we have travelled. Lao Tzu warned, "If you do not change direction, you may end up where you're heading." Today, this would be disastrous. The business of business as usual has been heading us toward socially, economically, and ecologically untenable conditions. Without a change in direction, we would be on the way to a world of increasing population pressure and spreading poverty; to growing social and political confrontation; to industry – and lifestyle-created climate change; to food and energy shortages; and to a worsening industrial, urban, and agricultural pollution of air, water, and soil. We would encounter more frequent and ever more devastating floods and tornadoes triggered by an unbalanced climate, and higher rates of mortality due to exposure to accumulated quantities of toxins in soil, air, and water.

We cannot afford to be further exposed to such conditions. We need fundamental change in the way we act, and that means change in the way we think. It means a change in the mindset that decides how we look at the world.

The Crucial Factor: Our Mindset

The Germans have an expression that's widely used, even in other languages. It is *Weltanschauung.* This is more than the view of the world based on science, or on any other doctrine or source. It is the ensemble of one's views of who he is, and what his world is, encompassing the associated values and beliefs.

Until recently, the mainstream in societies rich and poor, Western and Eastern, lived and acted with an outdated mindset. People were focused on material goods, personal wealth, and on ostentatious and wasteful "modern" lifestyles. They thought that the way they think and act is scientifically grounded. Individuals, they believed, have little or no influence on the way the world is going. And is if so, there is no need for individuals to feel responsible for the shape of the world. Life is a struggle for survival, and claiming otherwise is to indulge in wishful thinking. The "survival of the fittest" is a law that holds sway nature, and there is no sense in contesting it. In any event, an "invisible hand" (the hand of the free market) will sooner or later distribute the benefits.

The spread of this obsolete mindset, and the behaviors validated by it, have led to critical conditions in the ecological, social, and economic domains of the human world. It is high time to realize that this mindset is flawed: the world is not a giant mechanism operating according to rigid laws. In the real world the future is determined by how we perceive the world, and how we act in it. Today the future hinges on the way we act in our unstable and unsustainable social, economic, and ecological systems. How the world evolves is not decided by laws but by our values and behaviors. We know that the future will not be the same as the past, and not even as the present. It will be a different world that could be better or worse than the world today.

Fortunately, we could still shape our future into a better world. We need to adopt better thinking and acting—a better mindset. Here are three "clarion calls" that illustrate and embody the mindset we need.

Three Clarion Calls*

(1) The Call for Creativity and Diversity

A new way of thinking has become the necessary condition for responsible living and acting. Evolving it means fostering creativity in all people, in all parts of the world. Creativity is not a genetic but a cultural endowment of human beings. Culture and society change fast, while genes change slowly: no more than one half of

---

* These calls were first put forward in a Manifesto drafted by the present writer in collaboration with the Dalai Lama. The "*Manifesto on the Spirit of Planetary Consciousness* was adopted by the Club of Budapest at the Hungarian Academy of Sciences in 1996, at a meeting with the Dalai Lama and other leading personalities from science and spirituality.

one percent of the human genetic endowment is likely to alter in an entire century. Hence most of our genes date from the stone age or before; they could help us to live in the jungles of nature but not in the jungles of civilization. Today's economic, social, and technological environment is our own creation, and only the creativity of our mind—our culture, spirit, and consciousness—could enable us to cope with it. Genuine creativity does not remain paralyzed when faced with unusual and unexpected problems but confronts them openly, without prejudice. Cultivating it is a precondition of finding our way toward a globally interconnected society in which individuals, enterprises, states, and the whole family of peoples and nations could live together peacefully, cooperatively, and with mutual benefit.

Sustained diversity is another requirement of our age. Diversity is basic to all things in nature and in art: a symphony cannot be made of one tone or even played by one instrument; a painting must have many shapes and perhaps many colors; a garden is more beautiful if it contains flowers and plants of many different kinds. A multicellular organism cannot survive if it is reduced to one kind of cell—even sponges evolve cells with specialized functions. And more complex organisms have cells and organs of a great variety, with a great variety of mutually complementary and exquisitely coordinated functions. Cultural and spiritual diversity in the human world is just as essential as diversity in nature and in art. A human community must have members that are different from one another not only in age and sex; but also in personality, color, and creed. Only then could its members perform the tasks that each does best, and complement each other so that the whole formed by them could grow and evolve. The evolving global society would have great diversity, were it not for the unwanted and undesirable uniformity introduced through the domination of a handful of cultures and societies. Just as the diversity of nature is threatened by cultivating only one or a few varieties of crops and husbanding only a handful of species of animals, so the diversity of today's world is endangered by the domination of one, or at the most a few, varieties of cultures and civilizations.

The world of the twenty-first century will be viable only if it maintains essential elements of the diversity that has always hallmarked cultures, creeds, economic, social, and political orders, as well as ways of life. Sustaining diversity does not mean isolating peoples and cultures from one another: it calls for international and intercultural contact and communication with due respect for each other's differences, beliefs, lifestyles, and ambitions. Sustaining diversity also does not mean preserving inequality, for equality does not reside in uniformity, but in the recognition of the equal value and dignity of all peoples and cultures. Creating a diverse yet equitable and intercommunicating world calls for more than just paying lip-service to equality and just tolerating each other's differences. Letting others be what they want "as long as they stay in their corner of the world," and letting them do what they want "as long as they don't do it in my backyard" are well-meaning but inadequate attitudes. As the diverse organs in a body, diverse peoples and cultures need to work together to maintain the whole system in which they are a part, a system that is the human community in its planetary abode. In the last decade of the twentieth century, different nations and cultures must develop the compassion and the solidarity that could enable all of us to go beyond the stance of passive tolerance, to actively work with and complement each other.

(2) The Call for Responsibility

In the course of the twentieth century, people in many parts of the world have become conscious of their rights as well as of many persistent violations of them. This development is important, but in itself it is not enough. In the remaining years of this century we must also become conscious of the factor without which neither rights nor other values can be effectively safeguarded: our individual and collective responsibilities. We are not likely to grow into a peaceful and cooperative human family unless we become responsible social, economic, political, and cultural actors.

We human beings need more than food, water, and shelter; more even than remunerated work, self-esteem, social acceptance.

We also need something to live for: an ideal to achieve, a responsibility to accept. Since we are aware of the consequences of our actions, we can and must accept responsibility for them. Such responsibility goes deeper than many of us may think. In today's world all people, no matter where they live and what they do, have become responsible for their actions: as private individuals; citizens of a country; collaborators in business and the economy; members of the human community; and persons endowed with mind and consciousness. As individuals, we are responsible for seeking our interests in harmony with, and not at the expense of, the interests and well-being of others; responsible for condemning and averting any form of killing and brutality; responsible for not bringing more children into the world than we truly need and can support; and for respecting the right to life, development, and equal status and dignity of all the children, women, and men who inhabit the earth. As citizens of our country, we are responsible for demanding that our leaders beat swords into ploughshares and relate to other nations peacefully and in a spirit of cooperation; that they recognize the legitimate aspirations of all communities in the human family; and that they do not abuse sovereign powers to manipulate people and the environment for shortsighted and selfish ends. As collaborators in business and actors in the economy, we are responsible for ensuring that corporate objectives do not center uniquely on profit and growth, but include a concern that products and services respond to human needs and demands without harming people and impairing nature; that they do not serve destructive ends and unscrupulous designs; and that they respect the rights of all entrepreneurs and enterprises who compete fairly in the global marketplace.

As members of the human community, it is our responsibility to adopt a culture of nonviolence, solidarity, and economic, political, and social equality, promoting mutual understanding and respect among people and nations whether they are like us or different, demanding that all people everywhere should be empowered to respond to the challenges that face them with the material as well as the spiritual resources that are required for this unprecedented task. And as persons endowed with mind and consciousness, our responsibility is to encourage comprehension and appreciation for the excellence of the human spirit in all its manifestations, and for inspiring awe and wonder for a cosmos that brought forth life and consciousness and holds out the possibility of its continued evolution toward ever-higher levels of insight, understanding, love, and compassion.

(3) The Call for Planetary Consciousness

In most parts of the world the real potential of human beings is sadly underdeveloped. The way children are raised depresses their faculties for learning and creativity; and the way young people experience the struggle for material survival results in frustration and resentment. In adults this leads to a variety of compensatory, addictive, and compulsive behaviors. The result is the persistence of social and political oppression, economic warfare, cultural intolerance, crime, and disregard for the environment. Eliminating social and economic ills and frustrations calls for considerable socioeconomic development, and that is not possible without better education, information, and communication. These, however, are blocked by the absence of socioeconomic development so that a vicious cycle is produced: underdevelopment creates frustration, and frustration, giving rise to defective behaviors, blocks development.

This cycle must be broken at its point of greatest flexibility, and that is the development of the spirit and consciousness of human beings. Achieving this objective does not preempt the need for socioeconomic development with all its financial and technical resources, but calls for a parallel mission in the spiritual field. Unless people's spirit and consciousness evolves to the planetary dimension, the processes that stress the globalized society–nature system will intensify and create a shock wave that could jeopardize the entire transition toward a peaceful and cooperative global society. This would be a setback for humanity and a danger for everyone. Evolving human spirit and consciousness is the first vital cause shared by the whole of the human family.

In our world, static stability is an illusion, the only permanence is in sustainable change and transformation. There is a constant need to guide the evolution of our societies so as to avoid breakdowns and progress toward a world where all people can live in peace, freedom, and dignity. Such guidance does not come from teachers and schools, not even from political and business leaders, though their commitment and roles are important. Essentially and crucially, it comes from each person himself and herself. An individual endowed with planetary consciousness recognizes his or her role in the evolutionary process and acts responsibly in light of this perception. Each of us must start with himself or herself to evolve his or her consciousness to this planetary dimension; only then can we become responsible and effective agents of our society's change and transformation. *Planetary consciousness is the knowing as well as the feeling of the vital interdependence and essential oneness of humankind, and the conscious adoption of the ethics and the ethos that this entails. Its evolution is the new imperative of human survival on this planet.*

Answering the Calls: The Evolutionary Scenario

The unfolding of this evolutionary scenario is based on the emerging of a critical mass of people who are ready and willing to adopt a more responsible way of living. This, as we have already noted, calls for a fundamental change in thinking and acting of the critical mass.

Let us assume that this critical mass emerges in time. Then we could expect that the evolutionary scenario begins to unfold on this planet. The first development may be a united effort of the political and business leaders to safeguard the essential balances of the biosphere. They impose regulations that halt the dumping of CO_2 and other noxious waste products into the environment. As the results become evident, the regulations acquire wider following and remain effective even in the face of resistance by the conservative segments of business and political community. Public bodies and private enterprises comply with the new regulations partly because of the emerging insight that they are necessary, and partly because noncompliance entails heavy penalties.

Together with the regulations required to control climate change, governments redouble efforts to control the spread of the COVID-19 virus and other infectious diseases. Coordinated measures are put into practice to liberate the population from the threat of this and foreseeable further pandemics.

As worldwide coordinated measures begin to produce results, people begin to think differently. A sense of urgency to live and act responsibly is joined with a renewed commitment to build a more crisis-resistant world. With the new mindset, people come to see the planet as a living organism of which they are an integral part. This organism, they see, is under mounting stress; and in the absence of adequate measures, humans are becoming endangered species.

Young people, and sensitive and concerned individuals of all ages, realize that we are a vital element in a system of great complexity and high but increasingly unstable coherence. They realize that this system has been subverted, its evolution derailed. But they also realize that effective constraints are required to avert a downshift into critical and ultimately irreversible conditions that make society prone to a series of catastrophic breakdowns.

Before long, action prompted by fear of the consequences of inaction is replaced by action inspired by the perception of the positive possibilities. The stages of the evolutionary upshift are marked by a sequence of progressive developments:

- Seizing the opportunity to change and to transform, people pull together. There is growing support for public policies that exhibit a higher level of social and ecological responsibility. Funds and capital are beginning to be channeled from "defense" and "security" objectives and serving the affluent minority, to the basic needs of the great majority.
- Measures are progressively implemented to safeguard the environment, create an effective system of food and resource

distribution, and develop and put to work sustainable energy, transport, and agricultural technologies.

- More and more people gain access to food, jobs, and education. More and more enter interactive platforms on the internet and become active in spreading a dialogue that shapes people's thinking and affects their behavior. Their dialogue prompts them to join together, and together they discover more and more areas of common interest.

- Business leaders begin to change their operating modalities. They work toward creating a circular economy, responsible living on natural income rather than on spending its natural capital. Natural capital, they realize, consists of the riches we borrow from the earth, and these need to be repaid and not just used and discarded. When natural capital is depleted, the community or enterprise based on it goes bankrupt—incapable of maintaining its essential resource base. Drawing on natural income, on the other hand, means using renewable and recyclable energies and substances, and infinite or nearly infinite flow-energies such as wind, tide, and solar energy inflow. These resources have no expiration date, and the enterprises and communities based on them are indefinitely sustainable.

- A corresponding change takes place regarding the way other natural resources are used. The objective is no longer to optimize labor productivity (previously the principal goal of business companies), but to optimize resource productivity. Instead of squeezing the maximum out of every kilogram of matter and kilowatt of energy, the aspiration becomes to design resource-processing systems that require a minimum of nonrenewable materials and a maximum of quasi-infinitely available flow-energies.

- In time, the human community of the planet rises to the challenge of eliminating the specter of deepening crises leading to collapse. Technologically advanced economies pioneer sustainable resource processing systems to operate in the framework of sweeping social and political reforms.

When the new arrangements inspired by the new mindset take effect, people begin to live more responsibly on our home planet. A higher level of cooperation among communities, governments, and business companies ensures a more rational consumption of resources and a more efficient discarding of the wastes, thus reducing the human impact on the global ecosystem.

There is less stress within the social and political community itself. National, international, and intercultural animosity, ethnic conflict, racial oppression, economic inequity, and gender inequality diminish. A unilateral focus on satisfying national, corporate, and in-group interests is replaced by wider and less partisan politics: a politics dedicated to meeting the legitimate ends of two of the three Ps: People and Planet. The third P, Profit, is no longer an end in itself—it is a means to an end. The end is the creation of a thriving world for the third and most essential of the three Ps: People.

A Manifesto

Evolving the Self-Governance

of Human Societies

Preamble

We, the people of the planet, are at a crossroads in our evolutionary journey. We can choose to make great music and come into harmony as one, or continue on the current path of discordant chaos with the inevitable results—it all depends on whether we are willing to attune with humanity and the natural world.

In this singular moment in history, we must choose between the values of an old paradigm structured in competition, individualism, and materialism, or embrace a new mindset rooted in cooperation, the good of the whole, and a higher level of consciousness. We must look toward our indigenous roots of harmony and oneness with Gaia, with each other, and with Spirit, however we define it, for a balance of the feminine and the masculine and with an eye toward the well-being of future generations. Now is the time to choose a new way forward, one of peace and cooperation with the world around us, both inside and out.

To do any less, to ignore the current conditions of our time—the rampant fear, anger, exploitation, violence, social and financial inequality, and the rapid deterioration of our environment—is to continue on a path toward disaster, and to include ourselves potentially in the rapidly unfolding sixth extinction. If we see that our current universe is an aberration of the original vibration and it produces the cacophony of our times, it will be clear to us that our mindset is dangerously out of tune. Instead of cooperation within the band, leaders motivated by profit and power play their own instruments to their own tunes at the expense of the collective harmony that creates collective well-being.

Article 1

Our current governance framework reflects a musical score written from the perspective of a mechanistic and hostile universe, with a "winner-takes-all" system of values. Our world of nation-states competing for the perceived limited resources has obliterated our natural order of oneness with each other and the planet, and held us captive to a political machine dependent on strong borders, armaments, and violence. The illusion of separation has infiltrated our souls and allowed us to think that the music of life itself is a competition for volume at any cost, as if the musician who plays or sings the loudest is the most deserving at the expense of those who offer a more subtle or nuanced contribution. Now is the time for a collective song of reason and harmony to be sung within a new system of politics.

Article 2

In the emergent sacred governance in the global orchestra, fear dissolves into our natural state of love. This love is the simple resurgence of the golden rule that has existed from time immemorial in indigenous and holistic societies—in essence, to live in harmony with all fellow musicians. Our current belief that individual rights are sacrosanct is a mistake. What is good for the individual may not be good for the whole, whereas what is good for the whole orchestra is always good for every individual player. This is the moment when each individual must retune her instrument, so its sound will be in harmony with all the players in the planetary orchestra.

We ask that the leading figures of the self-governance of humankind become humble and committed conductors in service to the whole. Consider some specialized professions: doctors, teachers, engineers. There are high standards that must be met to call one-self a medical practitioner. It requires a commitment of years of study, examination, apprenticeship, and ongoing education. The Hippocratic oath, one of the oldest binding documents in history, continues to be held sacred by physicians. Let us envision a new day when the formerly ego-driven governors of nation-states are guided by a similar universal principle. Envisage the day when a sacred code of ethics is adopted, a code that enables leaders to be caretakers of a melodious, peace-loving, nature – and universe-loving society.

The new ethics calls to our evolving consciousness to embrace the journey toward a higher state of being. Our inherent altruistic nature will manifest in a safer and more compassionate world in which we will begin to play in tune with our deepest instincts, creating a world of cooperation with a win-win condition for all.

Article 3

Our evolution as a species embraces healing and reconciliation. To overcome the fear and greed that has driven political thought in the past, we strive to align ourselves with a frequency of vibration that brings us back into balance with the human spirit and with Mother Earth. We strive to expound our cultural and artistic impulses to honor the highest qualities of our endeavors within a political process anchored in sacred service to the good of the whole. The new stories of success are achieved by those who do the most good for society and the planet.

The emergent and restructured global governance system is based on a fortified United Nations with a World Federalist Charter, adhering to the Earth Charter and a new Constitution for the New Paradigm in Governance. Funded by a scaled system based on per capita income, the new world governance will provide relief for the poor and underserved. No longer will credit and privilege be given to those soloists who prefer the sound of their own voices at the expense of the harmony of the whole. At long last, the human symphony will resound with the frequency of our original sacred vibration.

Article 4

Behold the new mindset: a celebration and embodiment of goodwill and cooperation, based on the ideals of truth, beauty, and goodness. Sworn to sacred duty, our leaders will be trained and guided by a council of globally elected governors, highly skilled practitioners of right statesmanship. Unity is found in a new code of inclusion extending to all races, ethnicities, cultures, sexual identities, indigenous beliefs, and states of being. There is a place and a purpose for everyone in the new composition of universal citizenship.

Leaders who embody the new mindset understand the power of language and honor the responsibility of engaging in truthful communication on behalf of humanity and the planet. Instead of using human resources to manage divisiveness, new energy is generated from the cohesiveness of the governance process, allowing social justice and equal opportunity to contribute to an elevation of consciousness on an accelerated scale. Just as some countries have transformed by shifting the perspective from "me" to "we," all nations with the new mindset now surge forward with measurements of happiness. The new leaders pave the way for universal health care and education, allowing everyone to live a vital and balanced life. This new mindset is holistic—honoring the good of the whole as its first priority.

Article 5

Future generations will be born into a world honoring their unique form of being so we all could reach our highest expression: a world in peace, upheld by the World Federalist Constitution and maintained by the supervision of the United Nations, with an equitable distribution of power representing all peoples of the planet.

With leaders assuring the basic necessities for everyone, the rising tide of collective consciousness can be upheld by a guaranteed supply of safe and environmentally friendly food, clean drinking water, and fresh air for the entire human population, living in sustainable balance with nature. In this potentially noblest of professions, the politicians of the new mindset establish a sacred contract that each member of the human community can sign as a global citizen, activating a society based on direct democratic principles that guarantee equal voice and participation for all people. The contract is to include all parts and elements of society. It is to empower women to take their rightful place as equal partners leading the way to a better future drawing on the divine feminine. As such, love and unity is the message of the new anthems, issuing a call for political bodies to take action and heal the wounds inflicted in history.

Article 6

We hereby declare that each one of us has the responsibility and now also the opportunity to sing together and energize humanity to an ever-evolving higher state of consciousness. Alongside our politicians, we are all active participants in crafting this new vision of universal citizenship, engaged in transforming our current system that operates on a love for power into a new system impelled by the power of love.

Embodying our birthright to fulfill our sense of purpose for the good of the whole becomes a rite of passage. Within the new paradigm of politics, our innate goodness and respect for collective well-being empower all people and bring all people in the world to a deeper understanding of the universe as a living system in balance with human communities and their natural environment. The evolution of consciousness that comes as a result of political coherence economizes energy on all scales and brings our world into collective harmony. The enlightened politician is the conductor of the new symphony performed by all people in the human family.

Article 7

The music we will make together with the guidance of the new mindset will emit a vibration with a sound such as has never been heard before. This is the super coherent energy of a new field, beyond notions of right and wrong, left or right. It is the field we reach as our spiraling evolutionary path reclaims and embraces the values that drive humanity's heart: the common ground of truth, beauty, and goodness. In this collective vision we shall heal and prosper, preserving the sanctity of our planet and the sanity of our civilization for generations to come. Together we shall create the most beautiful performance ever attempted: life and governance fused into the synergistic energy of music, resonating and singing to all corners of the universe.

PART TWO

YOUR MISSION

CHAPTER 4
START ON THE WAY FORWARD

It's time to come down to brass tacks. These tacks, the fabled "buck," stop with every one of us. In the last count, it is up to us, to you and to me, whether the human family will shift up to a thriving world—or drift down to deepening crises and chaos. As a conscious and responsible member of humanity, you need to find the way forward—and enter on it. Here, to begin with, are the first steps to embark on this epochal journey.

The First Steps

Even the longest journey begins with the first step
Chinese proverb

Know Where You Want to Go

If you are to take the first steps toward responsible living on this planet, you must be clear about what it is that you want to accomplish. Your endeavor is to create a better world. But what is the nature of a better world?

Today's world, as we now know, is fundamentally unsustainable. Attempts to stabilize it in its current form are condemned to fail, even if they appear politically and technically "realistic." They may be realistic in the short term, but their realism evaporates when we look but a few years ahead. Coming back to where we were is not the answer. Restitutive measures may mitigate the worst of the socioeconomic gaps and the most visible forms of environmental degradation, but they do not overcome the problem—they do not change the system that gave rise to them.

Limiting CO_2 emissions, reducing income differentials, and safeguarding the quality of soil, air, and water are important goals, yet pursuing them is not doing something that is unconditionally good, only something that is less bad. In some cases, doing something less bad would not be good at all; it would extend the lifespan of the malfunctioning system and allow the problems engendered by it to intensify and become more and more intractable.

That doing something that is less bad is actually bad seems like a paradox. But this is the case because the more we delay reaching the point where the current social, political, and economic system is sufficiently destabilized to produce real change, the higher is the cost of implementing the needed change.

The global crisis in which we find ourselves is forcing us to change—it is forcing the hand of history. This is good, even fortunate. Because there are significant benefits brought by a crisis. Among other things, it creates solidarity—the will to pull together. This has been the experience during WWII, when people threatened by Hitler's armies did not quibble or fight each other, but joined force to face the shared danger. Countless acts of solidarity and heroism were born. But do we have to wait for a further social and political catastrophe to come up with the solidarity to pull together?

We cannot afford to wait. The "tried and tested" methods do not work, and trying them is a waste of time. They do not work even in the domain of the economy and in the sphere of technology. More and more economic growth and more and more sophisticated technologies are not the panacea they are believed to be. Economic growth, if not accompanied by the means and provisions of distributing its benefits, concentrates its payoffs on an ever smaller minority and exacerbates the gap between rich and poor. As the experience of recent years shows, a globalized economy focuses production, trade, finance, and communication in the hands of an ever smaller minority, and produces a backlash for the majority in the form of spreading unemployment, widening income gaps, and worsening environmental degradation. In this globalized economy, hundreds of millions are living at a higher material standard, but thousands of millions are pressed into abject poverty, subsisting in shantytowns and urban ghettos in the shadow of ostentatious affluence. The current forms of economic growth exclude an ever-growing segment of humanity and generate resentment, frustration, and violence.

In today's world, undifferentiated economic growth is both socially and politically explosive. It fuels resentment and revolt and provokes massive migration from the countryside to the cities, and from the poor to the affluent parts of the world. Under these conditions, organized crime—already a global enterprise—finds fertile ground; it engages in a whole gamut of activities, ranging from the control of information to prostitution, fake information, and traffic in arms, drugs, and human organs.

Technological advances without checks and balances in their application are likewise not the answer. Technology is a powerful tool, but it is a two-edged sword. Nuclear power offers a nearly unlimited supply of commercial energy, but the disposal of nuclear wastes and decommissioning aging reactors pose unsolved problems and constitute major threats. And the specter of nuclear meltdown remains on the horizon, whether it is due to technical accident or intentional terrorism.

Genetic engineering has a remarkable power to create bacteria-resistant and protein-rich plants, improved breeds of animals, vast supplies of animal proteins, and microorganisms that produce proteins and hormones and improve photosynthesis. But genetic engineering can also produce lethal biological weapons and pathogenic microorganisms, as well as abnormally aggressive animals and insects. Irresponsibly applied, genetic engineering reduces the diversity of species and harms the balance between populations in today's already destabilized ecologies.

Information technologies could create an interacting global community by linking people whatever their culture and ethnic or national origin, but the networks of communication, despite sophisticated preventive measures, remain open to violations of privacy and are prey to hackers. They remain open to illegitimate, illegal, and criminal uses, such as spy networks and arms trade. A growing number of information and communication networks are operating at the edge of legitimacy, serving underground interests such as sex, prostitution, pornography, and extortion.

What are we to do? If the usual prescriptions for addressing our problems no longer produce benefits, we have to replace them—we must change the system. The world we need to build must be a stable, sustainable, and equitable world. It cannot be a continuation of yesterday's world. This becomes clear when we extrapolate current trends into the future. Doing so we fail to get a sustainable world—we get a world that is not only prone to breakdown, but is inevitably breaking down. The future cannot be extrapolated; it must be freshly created.

In his March 1933 Inaugural Address, Franklin D. Roosevelt said that the only thing we have to fear is fear itself. This is wise advice. We are granted the chance to live up to it. We live at a time when we not only *need* to, but actually *can*, upshift up to a better world. Ours is not a time for fear; it is a time for action: for taking the next step on the way to a better world.

Rid Yourself of Obsolete Beliefs

What is it that we can do today to help create an upshift to a better world? The answer is evolving and implementing projects of global sustainability, and social and ecological fairness and livability. This, as we have been saying, hinges fundamentally on the way we think. We need a new mindset. How to go about creating it is outlined on the following pages.

Question irrational practices

In 2004, the Wisdom Council of the international think tank The Club of Budapest issued a declaration that questions a number of practices that hallmark the contemporary world. These practices are still widely followed, although they are paradoxical and even irrational.

The Wisdom Council asked,

Where is the wisdom in a system that:

- produces weapons that are more dangerous than the conflicts they are meant to solve?
- creates an overproduction of food, but fails to make it available to those who are hungry?
- allows half of the world's children to live in poverty and millions of them to suffer acute hunger?
- fails to appreciate and make use of the sensitivity, care, and sense of solidarity women bring to the family and the community, and could bring to politics and business?
- faces a gamut of tasks and challenges, yet puts more and more people out of work?
- gives full priority to maximizing the productivity of labor (even though millions are unemployed or underemployed) rather than improving the productivity of resources (notwithstanding that most natural resources are finite and many are scarce and nonrenewable)?
- requires unrelenting economic and financial growth just to function and not to crash?
- faces long-term structural and operational problems, yet bases its criteria of success on three-month accounting periods and the day-to-day fluctuation of stock exchanges?
- assesses social and economic progress in terms of the gross national product and leaves out of account the quality of people's lives and how well their basic needs are fulfilled?
- fights religious fundamentalism but enshrines "market fundamentalism" (the belief that unrestrained competition on the market can right all wrongs and solve all problems)?

- expects people in their private sphere to abide by the golden rule "treat others as you expect others to treat you," yet ignores this elementary principle of fairness and equity in politics (to treat other states as you expect other states to treat your state) as well as in business (to treat partners and competitors as you expect partners and competitors to treat your business)?

Such irrational beliefs and practices are held by many and perhaps most people in the modern world. Open your eyes and look at yourself. Do you think and act like this yourself?

Some beliefs you will have to question

The cult of efficiency. I must get the maximum out of every person, every machine, and every organization regardless of what is produced and whether or not it serves a useful purpose.

— Efficiency without regard to what is produced and whom it will benefit leads to mounting unemployment, catering to the demands of the rich without regard to the needs of the poor, and polarizes society into "monetized" and "traditional" sectors.

The technological imperative. Anything that can be done ought to be done. If it can be made or performed, it can be sold, and if it is sold, it is good for us and the economy.

— The technological imperative has resulted in a plethora of goods that people only think they need; some of them are not only not necessary, but actually harmful.

The newer is the better. Anything that is new is better than (almost) anything that we had even last year.

— That the new would always be better is simply not true. Often, the new is worse than the "old" that it replaced— more expensive, more wasteful, more damaging to health, and more polluting, alienating, or stressful.

Economic rationality is paramount. The value of everybody, including neighbors, partners, clients, and customers can be calculated in money. What I want is what everybody wants: to amass money. The rest is idle conversation and mere pretense.

— The reduction of everything and everybody to economic value may have seemed rational during epochs in which a great economic upswing turned all heads and pushed everything else into the background, but it is foolhardy at a time when people are beginning to rediscover deep-rooted social and spiritual values and to cultivate lifestyles of voluntary simplicity.

My country, right or wrong. Come what may, I owe unwavering allegiance only to one flag and one government.

— The chauvinistic assertion "my country, right or wrong" has played untold havoc both domestically and internationally, calling for people to fight for causes a new government later repudiates, to espouse the values and worldviews of a small group of political leaders, and to ignore the growing cultural, social, and economic ties that evolve among people in different parts of the globe. Yet there is nothing in the healthy human psychology that would forbid the expansion of loyalty above the level of the country one happened to have been born into. No individual needs to swear exclusive allegiance to one flag only. We can be loyal to several segments of society without being disloyal to any. We can be loyal to our community without giving up loyalty to our province, state, or region. We can be loyal to our region and feel at one with an entire culture and with the human family as a whole. As Europeans are English, Germans, French, Spanish, and Italians as well as Europeans, and as Americans are New Englanders, Texans, Southerners, and Pacific Northwesterners as well as Americans, so people in all parts of the world possess multiple identities and can develop multiple allegiances to them.

The future is none of my business. Why should I worry about the interest of the next generation? Every generation has had to look after itself, and the present and the next generations will have to do the same.

— Living without conscious forward planning—though it may have been fine in days of rapid growth when each new generation could ensure a good life for itself—is not a responsible option at a time of global shift, when the rationality of the past is in question and the rationality that will decide the life and well-being of future generations is not yet in place. Yet we already know that the decisions we make today will have a profound impact on the generations that come after us.

Outdated assumptions

The environment is an inexhaustible reservoir of resources.
The assumption that, for all practical intents and purposes, the environment provided by the planet's biosphere is an infinite source of resources and an infinite sink of wastes is persistent but irrational. Its origins go back to the archaic empires. It would hardly have occurred to the inhabitants of ancient Babylonia, Sumer, Egypt, India, or China that the world in which they lived could ever be exhausted of the basic necessities of life—edible plants, domestic animals, clean water, and breathable air—or fouled by dumping wastes and garbage in it. Nature appeared to be far too vast to be tainted, polluted, or defiled by what humans did in their settlements and on the lands that surround them.

The belief in an environment that is an inexhaustible fountain of resources gave rise to the way of relating to the environment that hallmarked most of recorded history. It originated nearly ten thousand years ago in the Levant. Until then, indigenous communities in Africa, Asia, and pre-Colombian America had a deep respect for the environment and used only as much as nature could regenerate. From the time of the so-called Neolithic Revolution (which saw the beginning of humans domesticating plants and animals for their own purposes), some communities began not only to exploit but to overexploit their environments. The Mycenean and Olmec civilizations and those of the Indus Valley are notable examples. In the Fertile Crescent, this practice produced historical consequences. There, at the cradle of Western civilization, humans were not content with the perennial rhythms and cycles of nature, but sought ways to harness nature for their own ends. The land, though hot and arid in spots, appeared amenable to exploitation. Yet there were problems even in the friendly environments. In some places, such as ancient Sumer, flash floods washed away irrigation channels and dams and left fields arid.

The assumption that the resources provided by the environment are inexhaustible made much of the Fertile Crescent of biblical times into the Middle East of today—a region with vast areas of arid, infertile land. But people could move on, colonizing new lands and exploiting fresh resources.

In the Nile Valley, the environment was more benign. Great rivers irrigated the land, brought in silt, and washed away wastes. The archaic civilizations were mostly riverine civilizations built on the shores not only of the Nile but of the Yellow River, the Tigris and Euphrates, the Ganges, and the Huang-Ho.

Today when we overexploit our environments we have nowhere left to go. In a globally extended industrial civilization that wields powerful technologies, the belief in the inexhaustibility of nature is irrational and dangerous. It gives reign to the overuse and thoughtless impairment of the resources of the planet and to the overload of the biosphere's self-regenerative capacities. Persistence in practices inspired by the belief in the inexhaustibility of the planet's resources leads to an impoverished and inhospitable environment incapable of supplying the resources required by an enormous and still growing population.

Nature is a giant mechanism.
A second obsolete assumption dates from the early modern age, a consequence of the Newtonian view of the world. The idea of nature as a giant mechanism was adapted to medieval technologies

53

such as watermills and windmills, pumps, mechanical clocks, and animal-drawn plows and carriages, but it fails when it comes to nuclear reactors, networked computers, and genetically engineered plants and microbes. Sophisticated technologies are not Newtonian machines, and they do not obey mechanical laws.

Yet the assumption that nature can be engineered like a machine has persisted to our day. The basic idea supporting this belief is that doing one thing can always be relied upon to lead to another thing—much as pressing a key on an old-fashioned typewriter always makes an arm lift and print the corresponding letter. On the modern computer, however, sophisticated programs interpret the information entered on the keyboard and decide the outcome.

The mechanistic concept works even less well when man-made technologies are interfaced with natural processes. The way a transplanted gene is expressed in a given species of plant is foreseeable for that plant, but it is not foreseeable when it comes to the interaction of that plant with other plants in its environment. The same gene that produces the foreseen and desired effect in the transgenic plant can produce unforeseen and undesirable effect in other species.

A "horizontal gene-transfer" is always a possibility, but its long-term consequences for the ecosystem are unpredictable. Such consequences have already proved disastrous for the integrity of nature and the productivity of agricultural lands in Asia, Africa, and Latin America.

Nonetheless, pre-pandemic industrial civilization persisted in treating both its technologies and its natural environment as mechanisms that can be engineered and reengineered. The result has been the progressive degradation of water, air, and soil, and the impairment of local and continental ecosystems.

Life is a struggle where only the fittest survive.
The third obsolete assumption dates from the nineteenth century. It is an entailment of Darwin's theory of natural selection. It claims

that in society, as in nature, only the fit survive. This means that if we want to survive, we have to be fit for the existential struggle— fit, in fact fitter, than our competitors. In the context of society, of course, fitness is not determined by the genes but is a cultural trait such as smartness, daring, ambition, and the will to pursue our ends without being blocked by the anticipation of their unintended consequences.

Transposing nineteenth-century Darwinism into the sphere of society is dangerous, as the "social Darwinism" of the Nazi ideology demonstrated. Hitler's regime justified the conquest of territories in the name of acquiring adequate *Lebensraum* (living space) and warranted the subjugation of other peoples in the name of racial fitness.

In our day, the consequences of social Darwinism have gone beyond armed aggression to the less evident but in some ways equally damaging merciless struggle of competitors in the market-place. All-out competition has produced widening gaps between rich and poor and concentrated wealth and power in the hands of the winners: corporate managers and international financiers. States and entire populations have been relegated to the role of clients and consumers and, if poor and powerless, were dismissed as marginal factors in the equations that determine success in the marketplace.

The market distributes the benefits

For most of the late twentieth and early twenty-first centuries, the financial community has operated on the assumption that, as the economists' doctrine maintains, "the market ensures optimal capital allocation through the efficient incorporation of all available and relevant information into share prices."

This assumption is widely shared by business people. It is a counterpart of the assumption that life is a struggle where the fittest survive. Unlike in nature—where insufficient fitness results in the extinction of the species—in society there is the market, a mechanism that distributes benefits and thus counteracts the processes of natural selection. The market is governed by what Adam Smith called the "invisible hand." It assures those who do well for themselves that they do well also for others. Wealth trickles down from the rich to the poor.

Belief in the invisible hand of the market is comforting, and this doctrine is often cited by the winners to avert reproach for not caring for the losers. Unfortunately, the doctrine of the free market leaves out of account the fact that markets are not entirely free. Yet it would distribute benefits only under conditions of perfect or near-perfect competition. There has never been perfect or even near-perfect competition in the world—the playing field is never level. Inevitably, it is distorted in favor of the winners and to the detriment of the losers. This produces the kind of skewed income distribution we have in the world: the poorest 40 percent have 3 percent of the global wealth, and the wealth of a few hundred billionaires equals the income of half of the world's population.

The more you consume the better you are.
This is the assumption that there is a correspondence between the size of a person's wallet and the personal worth of the owner of the wallet.

The equivalence of human worth with financial worth has been consciously fueled by the business community. Some companies covertly. if no longer openly, maintain that consumption itself is the ideal. After WWII, Victor Lebov, a US retailing analyst, could had put forward the consumerist philosophy in clear language. "Our enormously productive economy," he wrote, "demands that we make consumption our way of life, that we convert the buying and use of goods into rituals, that we seek our spiritual satisfaction, our ego satisfaction, in consumption. The economy needs things consumed, burned, worn out, replaced, and discarded at an ever-increasing rate." Such unashamed promulgation of unlimited consumption cannot be maintained today, but it can be implied. It is implied in advertising consumption that harms the world, and harms in the last count the consumer in the world.

The belief that consumption is good for the consumer and good for the economy has been a powerful engine of growth in the economy. This engine ignored the fact that failing to take into account that ever more people consuming ever more resources is not sustainable in a finite world. It leads to inequities, resentments, and stresses we have experienced in recent years.

Achieving economic ends justifies using military means.
The ancient Romans had a saying: if you aspire to peace, prepare for war. They aspired for peace and were ready for war. The Romans had a worldwide empire, with rebellious peoples within and barbarian tribes at the periphery. Maintaining it required a constant exercise of military power.

The belief in using military means to achieve our ends persists in the modern world. Most people agree that if our economy is threatened, we have good reason to "call in the Marines." To aspire to peace, we have to be ready for war.

As the wars fought for the control first of coal and then oil demonstrated, recourse to armed intervention was considered legitimate when economic and territorial interests required it. Yet this belief is counter-functional in a world where local conflicts can escalate to global confrontation, and the parties to the conflict have access to weapons that can harm and destroy not just the presumed enemy, but all parties to the conflict. Using military means to achieve economic ends is a high-risk and irrational option in the twenty-first century.

Rethink even long-standing convictions.
In addition to the beliefs and assumptions we have just reviewed, a number of less entrenched but still widespread convictions are badly in need of being discarded. Here are a few:

- *Everything is reversible*
  The current crisis, as other epidemics and crises before it, is but a temporary perturbation after which everything

goes back to normal. All we need to do is manage the problems with tried and tested methods. Business as unusual has evolved out of business as usual, and sooner or later it will reverse back into it.

- *Everyone is unique and separate*
  We are all unique and separate individuals enclosed in our skin and pursuing our own interests. The same as our country, we have only ourselves to rely on; everyone else is either friend or foe, at best linked to us by ties of mutual—even if provisional and temporary—interest.

- *Order calls for hierarchy*
  Order in society can only be achieved by rules and laws and their proper enforcement, and this requires a chain of command recognized and obeyed by all. A few people on top (usually males) make up the rules, legislate the laws, give the orders, and ensure compliance with them. Everyone else is to obey the rules and take his or her place within the political hierarchy.

- *The ideology of Westphalia*
  The legal conventions coming into force at the end of WWI with the Peace of Westphalia conferred on nation-states the "inalienable right" to have an independent government, internationally recognized boundaries, a national currency and a national army, diplomatic relations with other states, and action free from fetters within their own borders. The formally constituted nation-state became the sole political authority, the only entity possessing legal and political sovereignty.

These convictions are steering us the wrong way.
— No experience of shocks and crises will change our values and behaviors if we remain convinced that the problems we encounter are temporary disturbances in an unchanging and perhaps unchangeable status quo.

— Seeing ourselves as separate from the rest of the world could convert natural impulses to seek our own advantage into a win-lose struggle to achieve our own aims regardless of the cost to others. This is a shortsighted strategy to follow for individuals as well as for states and businesses.

— Male-dominated hierarchies have worked in the army and also in the organized church, but they have not proven effective in business and in education. Successful managers and wise deans have learned the advantages of cooperative structures, but social and political institutions have continued to operate in a hierarchical mode. As a result, both state administrations and schools and universities have tended to be slow, their workings cumbersome and inefficient.

— Admitting nothing but your own nation as the focus of your allegiance is a mistaken form of patriotism. It often leads to chauvinism and xenophobia, and to such excesses as "the war on terror" in the name of national security, and "ethnic cleansing" ordered by dictators. But even democratic states have turned out to be inadequate to cope with the problems they face; they are too big in some respects and too small in others. On the one hand, decisions that touch people's lives—whether through education, employment, law and order, or civil liberties—require decision-making that is closer to the grassroots than the majority of today's national governments. On the other hand, decision-making in the economic and environmental spheres requires a level of control and competence that exceeds the effective scope even of the most powerful nation-states.

Review Your Aspirations
If the human community is not to fall back into the inequitable, dangerous, and downright irrational practices that still hold sway in today's world, you and other responsible and awakened people need to review and adapt your aspirations.

You can start by focusing on the paramount task of looking for, and actively seeking, a better balance between needs and demands. This balance has not been achieved in recent years, and it will not be achieved until there is a fundamental change in the way people manage the resources they require.

The current distribution of the benefits of economic activity is unfair and unsustainable. The rich are still getting richer, and the poor, poorer. The population of the poor has been growing faster than the population of the relatively affluent. If this trend were to continue, by the middle of this century over 90 percent of the more than nine billion people who will then live on earth would live in poverty.

When it comes to resource use, the issue is not numbers, but quality. It is not how *many* people use the planet's resources, but *how well* they use them. Our world has enough, as Gandhi said, to provide for all people's *need*, but not enough to provide even for one man's *greed*.

In the industrialized parts of the world, greed is dominant. In the name of freedom and laissez-faire capitalism, irresponsible values and beliefs give reign to selfishness and ostentation. Living by the dominant values and aspirations entails the excessive consumption of nonrenewable resources and the excessive production of waste. Those who "enjoy" the modern world's living standard use 80 percent of the world's energy and raw materials and contribute the lion's share of its pollution.

Irresponsibility and greed show up also in people's eating habits. The affluent consume an unsustainable amount of processed foods and expensive items such as red meat. The planet's entire grain harvest would not be enough to feed the cattle that would be needed if all people in the world would adopt such eating habits.

Those who can afford it use a disproportionate share of commercial energy as well. They heat their homes with inefficient gas-powered heaters or electric radiators and leave air conditioners running all day. Until recently, they were driving gas-guzzling vans, pick-up trucks, and sport utility vehicles; in preferences to hybrids and electric vehicles.

It merits remarking that, even though the affluent are using an excessive share of the planet's resources, they do not achieve by that a better quality of life. The modern ideal of luxury is flawed. The advertising world's paradigm of luxury—lounging by the pool, smoking a cigarette, sipping a daiquiri, munching on a hamburger, and talking on the cell phone—backfires. This lifestyle is neither healthy nor satisfying. It can lead to a number of chronic diseases, including skin cancer, lung cancer, cirrhosis of the liver, high cholesterol, and brain damage. In the last count, leading a life of luxury is not much of an improvement over working in a high-pressure job, taking smoke breaks every hour, having a drink after work to relax, and falling asleep in front of the television.

There are better ways for you to live your life. There are meaningful and important tasks you can achieve whatever your occupation and ambition. There are scores of healthy and rewarding ways to spend one's leisure time. Helping neighbors, creating a better community, visiting sites of natural, historical, or cultural interest, hiking, swimming, biking, reading, listening to music, or taking an interest in literature and culture are satisfying pursuits that do not involve a high level of material and energy consumption and do not require a great deal of money. They are healthier for you and easier on the environment than striving to satisfy the mainstream ideal of luxury.

Better aspirations on your part would give rise to behaviors that free a significant portion of the planet's resources for human consumption. For example, it takes 190 square meters of land and 105,000 liters of water to produce one kilogram of grain-fed feedlot beef, but to produce one kilogram of soybeans takes only 16 square meters of land and 9,000 liters of water. On the same amount of land where farmers catering to the preferences of the rich have been producing one kilogram of beef, they could produce almost 12 kilograms of soybeans or 8.6 kilograms of corn. It would save

96,000 liters of water to choose soybeans and 92,500 liters to choose to go with planting corn.

Eating fresh produce, living closer to nature, using public transportation, and walking on foot instead of sitting in cars are healthier than eating red meat and junk food, and sitting in cars in overcrowded cities and jammed highways. Given the rapid erosion of agricultural lands and the threatening water squeeze, adopting more responsive aspirations on the part of a critical mass is essential for the health of the planet, as well as for the health of people on the planet.

The responsible kind of aspirations need to be guided by a morality that is fair to all. The applicable moral principle is that whatever is moral is good for everyone. A morality based on this principle is new to the modern world, but was shared in traditional societies. The great religions have been setting the norms of public morality and they recognized the need for fairness in the distribution of values and resources. The Ten Commandments of Jews and Christians, the provisions for the faithful in Islam, and the Rules of Right Livelihood of the Buddhists are examples. But in recent times, the rise of science reduced the power of religious injunctions to regulate social behavior. Scientists, with some notable exceptions, did not come up with principles that would provide a basis for a universally acceptable moral code. Saint-Simon in the late 1700s, Auguste Comte in the early 1800s, and Émile Durkheim in the late 1800s and early 1900s tried to develop a "positive" scientific observation – and experiment-based ethic, but their initiatives conflicted with the science community's commitment to value neutrality. It was not picked up by the mainstream in society.

In the late twentieth century, ever more scientists joined spiritual leaders in recognizing the need for principles that suggest a universal code for moral behavior. The Union of Concerned Scientists, an organization of leading scientists, issued a statement in 1993 signed by 1,670 scientists from seventy countries including more than a hundred Nobel laureates. The ethic we need, said the scientists, "must motivate a great movement convincing reluctant leaders and reluctant governments and reluctant peoples themselves to effect the needed changes." The signatories noted the human responsibility for caring for the planet and warned that "a great change in our stewardship of the Earth and the life on it is required if vast human misery is to be avoided and our global home ... is not to be irretrievably mutilated."

In the last few decades, vast human misery has not been avoided, and we are not sure whether or not our global home has been irretrievably mutilated. In any event, the human community needs a trustworthy and fair moral code that takes into account everyone's health and well-being, not only the health and well-being of the rich and powerful. Your aspirations, and those of a critical mass, need to be reviewed and if necessary modified, so as to be adapted to the requirements for a healthy and sustainable global home.

Value Diversity!

Updating your obsolete beliefs and assumptions, reviewing and perhaps reconsidering your aspirations, and supporting the cultures emerging at the periphery are essential steps, but there are also other steps you need to take. You must come to value diversity and allow for the active appreciation of cultures that may be very different from yours.

Living in harmony with different peoples is a challenge. They think differently, have different values and ideals, and different lifeways. Yet we must live in harmony with them if this planet is not to be an arena of growing conflicts and breakdowns.

The difficulty of living in harmony with other peoples and cultures is not to be underestimated. Most people in the Western world think that everybody is like them and wants to live like them— claiming otherwise, they believe, is but sophistry and pretense. In the final count, everybody wants the same thing: money, power, sex, and to have a good time.

This belief does not check with reality. While it is true that the values of Western technological society shape many people's behavior, a great deal of diversity remains underneath. There is diversity in the way people view themselves, their society and nature, and conceive of liberty and justice. Disregarding, or even underestimating, the diversity of the cultures of the world has produced disastrous consequences in the past. It produced conflict escalating into bloodbaths in the Balkans, Ireland, the Middle East, the Arab world, sub-Saharan Africa, Latin America, the Indian subcontinent, as well as in Southeast Asia. Even international terrorism, as distinct from aggression and violence due to current or inherited prejudice and intolerance, has a cultural basis. It is fed by intense hatred, resentment, and the desire for vengeance.

Whether in the Balkans, in the Middle East, or elsewhere in the world, there is a need for better understanding cultural differences among people and societies. The spread of the internet, of MacDonaldism, of Coca-Colonization, and of modern consumer culture did not eliminate cultural diversity—it only masks it. Underneath the surface, the diversity of the human world remains undiminished.

Here is a brief overview of the cultural diversity we must not just tolerate, but value.

A Pocket Atlas of Cultural Diversity

- In the southern half of the Americas, an aggressive brand of cultural nationalism has been emerging. Latin Americans resent their dependence on North America and resent being receivers rather than producers of the culture that shapes their societies. As part of their cultural heritage, they have a deep inherent spirituality that contrasts with the pragmatism of the culture of the United States.

  There are transcendentalist elements of the indigenous Latin American culture that date back to the fifteenth century. The Catholic scholasticism of the European Middle

Ages was not merely a monastic philosophy—it was a spiritual system that governed every aspect of life. Subservience to ecclesiastical authority, like subservience to God and king, became axiomatic in everyday morality. Even when the colonial epoch drew to a close, there was no accommodation in Latin America between the scholastic legacy and Anglo-Saxon pragmatism. The businesslike mentality of the Anglo-Saxon world has never taken hold in the Southern part of the American continent.

- Western cultural domination is an agonizing issue for Arabs; they perceive it as an element of industrialized-country hegemony penetrating their countries. The Arab countries find themselves at the passive end of a dialogue that links them mainly with the culture of Western Europe and North America. The militant fundamentalism that has been emerging in the Arab world has been an expression of the resentment felt by Arab politicians, business leaders, and intellectuals in regard to the foreign domination emerging in this dialogue. In the Muslim culture as a whole, transcendentalism has been combined with monotheism, and in Sufism, it acquired a mystical streak.

- Mysticism has also been prevalent in the indigenous cultures of black Africa. These cultures have always been spiritualistic and animistic, and these features have not been eliminated in the traditional sectors of the population by the zeal of Christian missionaries, nor have they been overcome by the marketing propaganda of transnational corporations.

- India and the countries of South Asia have had prolonged contact with British culture, but despite their admiration and assimilation of many of its traits, they are intent on protecting their own millennia-old cultural heritage as shown by their deep admiration for Gandhi, Aurobindo, and their numerous spiritual leaders.

- Admiration mixed with fear has been a hallmark of the cultures of the young nations of sub-Saharan Africa. Though avid consumers of industrial culture, African leaders have been intent on fortifying the native cultural heritage. While the poor segment of the population remains steeped in traditional beliefs and ways of life, a small—and lately shrinking—elite of intellectuals has been searching for the roots of African identity and seeking the political power to reinforce it.

- Though in a different form, transcendentalism is also a feature of the Hindu and Buddhist cultures of the Indian subcontinent. It focuses people's attention on spiritual matters and functions as a counterweight to the rising materialism and consumerism of the "modernized" sector.

- The Oriental mind has conserved many aspects of its cultural heritage. The great cultural circle that radiated from China during the past millennium was shaped by the naturalism of Lao Tzu, the social discipline of Confucius, and the Buddha's quest for personal enlightenment. In the twentieth century, these cultural origins branched in different directions, giving rise to the orthodox culture of Mao's Yan'an, the pragmatic culture of Hong Kong's Kong-Tai, and the mix of naturalism, Confucianism, and Buddhism that still characterizes the culture of Japan. Asian cultures became "modernized" but not Westernized. Oriental work habits, group loyalties, and lifestyles have remained culture-specific and different from those of Europe and North America.

- In modern-day Russia, historical experience has made for a profound ambivalence regarding Western culture. There is persistent admiration for the achievements of the West in technology as well as in high culture, but there is also a persistent demand to ensure that foreign influences do not overwhelm Russia's own cultural heritage.

- On the European continent, the same as in North America, materialistic individualism and pragmatism are still dominant, but they are not uniform and not monolithic. They are laced with a rising appreciation, although we share the same continent, but we have not become the same ourselves.

New elements have been added today to the diversity of the modern world; elements that heighten, and not flatten, its diversity. Now we not only share the modern consumer culture with its shopping malls and multiple enticements, not only get on the same internet, watch the same programs on television, eat the same foods and dress in the same ways, we also face common threats to our health, experience common suffering from extreme weather, and experience increasing uncertainties in regard to our jobs and means of livelihood. Faced with these issues, people do not all respond in the same way: they respond in light of their own values and worldview. Contemporary peoples are diverse crews in the same boat. Their boats are becoming increasingly the same, but the crews themselves are not.

Your understanding and respecting the diversity of the values and worldviews that shape the world is essential if you are to make an effective contribution toward building a world where diverse people live in peace and cooperate for the common good.

The Ten Commandments of Living in a World of Diversity

1. Live in ways that meet your needs and allow you to pursue your objectives without detracting from the chances of other people to meet their needs and pursue their objectives.
2. Live in ways that respect the right to life and to economic and cultural development of all people, wherever they live and whatever their ethnic origin, sex, citizenship, station in life, and belief system.
3. Live in ways that safeguard the right to live in a viable environment of all the living things that inhabit the earth.
4. Pursue happiness, freedom, and personal fulfillment in harmony with the integrity of nature and with consideration of the related pursuits of others around you.
5. Require of your political leaders that they relate to other peoples peacefully and in a spirit of cooperation, recognizing

the legitimate aspirations for a better life and a healthy environment of all members of the human family.

6. Require of your business leaders that they accept responsibility not only for the owners and shareholders of their enterprises, but for all its stakeholders. They are to produce goods and offer services that satisfy all legitimate demand without impairing nature and reducing the opportunities of small enterprises and poor economies to compete in the marketplace.

7. Require public media to provide a constant stream of reliable information on basic trends and crucial processes in order to enable people to reach informed decisions on issues that affect their health, prosperity, and future.

8. Make room in your life to help those less privileged than yourself to live a life of dignity, free from the struggles and humiliations of abject poverty.

9. Work with like-minded people to preserve or restore the essential balances of the environment in your neighborhood, and if you can, in other parts of the world as well.

10. Encourage young people and open-minded people of all ages to evolve the spirit that could empower them to make ethical decisions on issues that decide their future, and the future of their families, friends, and children.

Curate the Emerging Cultures

We need to shift up from a world of unsustainable and crisis-prone ecology and inequitable economic and technological globalization to a sustainable and humane world. For such a shift to occur, a culture of sustainability and cooperation must evolve. Such cultures are already emerging on the horizon.

According to a simple but meaningful definition, a culture is the ensemble of the values, worldviews, and aspirations that characterize a group of people and distinguish it from others. Functionally integrated societies have their own cultures, and their cultures are seldom monolithic. There is usually a mainstream culture at the center and a number of alternative cultures on the periphery.

There is a culture emerging at the periphery that holds particular promise. It is made up of people who are rethinking their preferences, priorities, values, and behaviors, and are ready to shift from the ideal of consumption based on quantity toward quality defined by environmental friendliness, sustainability, and the ethics of production and use. They seek to, and in some instances already do, replace matter – and energy-wasteful technologies and practices with lifestyles hallmarked by voluntary simplicity and the search for coherence with people and nature. As this nascent culture moves toward the center, those who embrace it need to bypass the negative characteristics of New Age cultures, such as antisocial activities, promiscuous sex, and isolation. They need to be united by the ideal of living a natural and responsible life, rejecting the heartless impersonality and mindless destructiveness of the mainstream culture.

The experience of inner city deprivation and violence, and of the impotence of police measures to cope, are prompting young people, to embrace a culture of peace and nonviolence. Such a culture is emerging on the periphery, and it needs conscious attention and curating. This is an important and practicable task. A better culture is being born, and could grow.

California's Institute of Noetic Sciences (IONS) summed up the developments that mark the birth of new cultures in reference to a set of simultaneously occurring shifts:

The shift from competition to reconciliation and partnership: a change from relationships, organizational models, and societal strategies based on competition to relationships and models based on principles of healing, reconciliation, forgiveness, and male–female partnership.

The shift from greed and scarcity to sufficiency and caring: a change in values, perspectives, and approaches from the traditional self-centered and greedy mode toward a sense of the sufficient and the interpersonal concern of caring.

The shift from outer to inner authority: a change from reliance on outer sources of "authority" to inner sources of "knowing."

The shift from separation to wholeness: a recognition of the wholeness and interconnectedness of all aspects of reality.

The shift from mechanistic to living systems: a shift of attention from models of organizations based on mechanistic systems to perspectives and approaches rooted in the principles that inform the world of the living.

The shift from organizational fragmentation to coherent integration: a shift from disintegrative, fragmented organizations with parts set against each other to goals and structures integrated, so they serve both those who participate in the organizations and those around them.

A rapid shift of the emerging cultures toward the center may seem overly optimistic; in normal circumstances it would be utopian. But the global crises create instability at the center, and this instability allows peripheral cultures to grow both in numbers and in influence. Butterfly effects surface—minute fluctuations can trigger major storms. This means that, as famous anthropologist Margaret Mead aptly remarked, even a small group of people can change the world.

There are many ways you can support the rise of the emerging cultures toward the center. The following ways are particularly promising:

The way through self-exploration

People who practice meditation or engage in intense prayer can contribute to the needed culture shift by exploring their own intuitions, values, and motivations. They find elements in the deep dimension of their consciousness that invite them toward harmony with people and planet. Astronauts who had the privilege of traveling in space and viewing Earth in its living splendor discovered these elements in their own psyche. The intense tie they experienced to their planetary home changed their life, and it seems to persist for the rest of their life.

Not only astronauts who could behold our planet from outer space, but ordinary people here on earth can experience our oneness with the planet. There are various notable experiences that produce this shift in our mindset—the experience known as the NDE (near-death experience) is a prime example. People who return from the portals of death see themselves and the world in a new light. They have a fresh appreciation of life and a deep reverence for nature. They develop humanitarian and ecological concerns. They realize that we cannot do anything to others without doing it to ourselves.

Evidently, not everyone can be expected to engage in deep meditation, travel in space, or have near-death experiences. Yet everyone can contribute to this shift because, as psychiatrist Stanislav Grof pointed out, everyone can enter altered states of consciousness. In these states, we experience space – and time-transcending ties to other people, and to the world around us. Grof noted that he has yet to meet a single person, no matter what his or her educational background, IQ, and profession, who would have had altered-state experiences and continued to subscribe to the fragmented materialist view of the world.

The way through the experience of beauty and significance
Self-exploration through altered states of consciousness is not the only way you can promote the advance of a timely alternative culture. The experience of beauty and significance is another way. This experience can be catalyzed by nature, as well as by human artifacts.

The beneficial effects of contact with nature have been well known to traditional cultures. They are rediscovered and activated today under the label of "nature therapy. " In Japan, for example, many people practice "forest bathing": *Shin-rin yoku*. Forest bathing calls for going into a forest and feeling yourself become one with it—hearing the wind rustle through the leaves, sensing the play of light on the surface of a pond, feeling yourself floating with the clouds in the sky. Even to be in a forest proves healing. The sounds of nature sooth the nerves and calm the spirit.

The healing power of the nature experience is enhanced by direct bodily contact. This means embracing the trunk of a tree, or walking barefoot on unpaved ground. Such contact intensifies the experience and makes one *feel* the world. It encourages the insight that we are in everything, and everything is in us.

The experience of beauty and significance can be catalyzed by human artifacts as well. Philosophers and psychologists call this the "aesthetic experience." In moments of inspiration, painters, poets, musicians, writers, and other visionary and creative individuals experience oneness, solidarity, and a deep love for all things. The artifacts they create express this experience and catalyze it in others. Regardless of the particular mode in which the aesthetic experience is expressed, it connects the experiencer with people, with nature, and with the world at large. It lends credence to our intuitions of belonging to something larger than ourselves—to something sacred. According to psychologist William James, the experience of belonging is also conveyed by intense prayer, and contemporary psychotherapists find that it is likewise conveyed by the practice of mindfulness and meditation.

The experience of beauty and significance catalyzed by nature and by artifacts is not difficult to achieve. This experience is not limited to forests and mountains, or to museums, galleries, and concert halls. In one form or another, art is present throughout the human world. It shapes cities through architecture and urban design, enters our feelings through music and dance, and conveys deep insight through literature. No matter in what form they are expressed, experiences of oneness are powerful ways of curating the rise of the emerging cultures.

The way through science

Innovations in science (unless they have immediate technological, economic, or social applications) are not immediately known to a wide layer of the population. Scientists use esoteric language and complex mathematics; their treatises are neither accessible nor understandable beyond their disciplinary fields. The result is that the general public is poorly informed about the latest revolutionary advances of scientific research. This fails to exploit an important resource for recovering the timeless insights of the human mind.

There are scientists and science-minded thinkers who are aware of the importance of filling the communications gap and have formulated insights coming to light in the sciences in everyday terms. Paul Davies, Deepak Chopra, Gary Zukav, and Gregg Braden are among those who have contributed to the dissemination of key scientific insights to a wide public. They have written popular science books and held widely accessible lectures and podcasts on current advances in science.

Science, it appears, is evolving toward a holistic understanding of the human being and the world. The classical ideas of Newton, Darwin, and Freud have been overtaken by new discoveries. In light of the contemporary sciences, matter, life, and mind are consistent elements within an overall system of great complexity yet coherent and harmonious design. The biosphere is born within the womb of the universe, and mind and consciousness are born in the womb of the biosphere. Nothing is independent of any other thing. Our bodies are part of the biosphere, and they resonate with the web of life on this planet. Our minds are part of our bodies, and they are in touch with other minds near and far.

The real world is not limited to what we perceive with our senses. In extends beyond the here and now. Whatever happens here also happens throughout space and time, and whatever happens in this moment has grown out of what has happened at times past and is the womb of what will happen in the future.

Looking into our own deeper selves, and experiencing works of art and other forms of human creativity, give credence to our ties to each other to the world around us. These ties are confirmed by cutting-edge science. If an effective upshift is to happen in time, this insight needs to be brought to the consciousness of a critical mass among the people.

Chapter 5

Be the Change you Wish to see in the World

A better world can be created—a world in which humankind can thrive. To create such a world, major changes have to be made. You, and other conscious and responsible persons, can be effective agents in catalyzing these changes. To become such agents, you need to align yourself with evolution in nature and promote its influence in the human sphere.

As we said, evolution drives toward wholeness and harmony within and between the evolving entities. It is nonlinear, unfolding through periodic crises. Yet coherence is growing even through the disruptions—the crises are temporary fluctuations on the surface of a larger evolutionary wave. Evolution creates, and notwithstanding the fluctuations keeps creating, complex and coherent systems: entities that are coherent with each other and with the world around them.

You are part of the evolution that unfolds on this planet. You are a conscious being who can consciously decide to tune himself and herself to this evolution. You have a sensitive brain and nervous system, and a uniquely articulate consciousness. You have not been using them adequately. You allowed selfishness and narrow horizons to dominate thinking in the world. They produced chaos and conflict, fed by intolerance and violence. You need to change yourself, so you could change the world.

Gandhi's saying is the key. *Be* the change you wish to see in the world—then others will resonate with you and will themselves change.

How You Can Be the Change

If you are to be the change you wish to see in the world, you need to change yourself. This calls for changing your mindset.

Change Your Mindset

Living consciously and responsibly in today's world calls for a more up-to-date mode of thinking and experiencing the world. This is a precondition of "going with the force" and shifting up to a better world.

Can you acquire a more up-to-date mindset? This is something only you can answer. You need to ask yourself some honest questions. How do you relate to people, society, and nature? How do your actions affect others and the world around us? There are many ways of thinking and not all of them have positive effect. You may have to shift your thinking about your relations to people and to nature.

The upshift we need is from fragmented and fragmenting, to holistic and wholeness-promoting thinking. Evolution drives toward wholeness, and your mindset needs to reflect the evolutionary trend. You need to think in terms of wholes—in terms of coherent and integral entities that are more than the sum of their parts. You must realize that the things and events that meet your eyes may be fragments, manifestations of larger systems that are coherent and whole. What you see in the world is largely incoherent—even chaotic—yet, as the new sciences testify, the world around them is intrinsically coherent.

To discover wholeness in the world, you need to look at the world through holistic lenses. They give you a view that is closer to reality than the fragmenting lenses you and most people are still using.

The holistic view of the world is a basic element of the mindset we need in the world. The fragmented, mechanistic view of the

75

world supported self-promoting fragmented actions in the world. It broke apart the integrity of the social, economic, and political world, of the entire the system of life on the planet. This you need to recognize. The human community, the same as all life on earth, is healthy when it is whole, and it is sick when it is fragmented. We cannot reduce a whole system to its parts without losing the very features that make it whole—and make it sound and healthy.

The sciences of life are clear on this point. The web of life on the planet cannot be dissembled into separate parts. Life is one, and when it is cared for as one, it persists and thrives. When it is segmented into its parts, it decays and degenerates. We are healthy when we are whole, and life on the planet is healthy when it is whole. Intuitively and instinctively we have known this, but we have buried our insight under a stressful and ever more desperate scramble for money and power—for what we perceive to be our own interests.

We have disregarded the whole for the sake of the part—for "my" part. We have developed amazing energy and information technologies but applied them single-mindedly to achieve our immediate ends, without regard for the consequences for others and for nature. We have made our technologies respond to our superficial wants, but allowed them to neglect our real needs. We have created an unsustainable world, prone to crisis and breakdown.

We need to adopt a new, holistic mindset. Given the bright legacy of the global crisis, you can live up to this challenge. You can adopt the mindset that enables you to see yourself as an intrinsic part of the grand sweep of evolution that governs existence on this planet. That practical implication of this mindset can be summed up in a single sentence: _whatever is truly good for the whole is bound to be good for all its parts._

The holistic mindset has been a standard feature of the mindset of traditional societies. People "knew" that what is individually good for them and for the other members of their community is that which is good for the entire community. In modern societies, this insight has been abandoned. Business and political leaders extol the individual and single-mindedly serve his or her interests.

The operative principle is contrary to the holistic principle: what is good for the part is assumed to be good for the whole. Political and business leaders pursue the good of their own country or company with the tacit assumption that it will work out to be good also for humanity. This, however, has seldom proven to be the case.

Operating on the basis of this "unholy" principle often produces strongly negative consequences for the larger community in which the given state or enterprise is embedded. It fractures that community, producing conflict, competition, and stress.

The unholy principle is the ideology of individual nation – or enterprise-centered politics. It came to the fore in the US congressional testimony of Charlie Wilson, then president of General Motors. Speaking at a congressional hearing, he said, "What is good for General Motors is good for the country." Few people have been contesting this philosophy. Even Kennedy said that "a rising tide lifts all boats." We would need to add that a rising tide does lift leaking boats… There is a lemma to the validity of the principle that "wealth trickles down" from the rich to the poor. In reality, without care for the community of human life on the planet, the rich become richer, and the poor, poorer.

That the immediate interests of the individual do not necessarily translate into the interests of the community in which that individual is embedded holds true for states and nations. Putting "America first" has been a shortsighted policy. It had motivated the Trump administration's withdrawal from the Paris Accord on Climate Change, its rejection of multilateral cooperation with other countries in the Western Hemisphere, and its engaging in trade wars with China and stopping support for international bodies such as UNESCO and the World Health Organization.

The same philosophy was behind Hitler's proclamation, *Deutschland ueber Alles* ("Germany above all"). This principle legitimized Germany's invasion of Poland, annexation of Austria, and the world-domination ambitions of the Nazi regime.

Sooner or later, the unholy principle produces a backlash. At the end of WWII, it created disastrous conditions for Germany, and had it been pursued by the US administration, it would have produced disastrous consequences for America itself.

Upshift Your Spirituality

This brings us to the next point in this review of ways you can become the change you wish to see in the world. This point raises a perennial, and today again highly topical question. Is there an actual need for spirituality in the modern world?

Aristotle said that we are social beings, and modern consciousness research confirms this. It affirms that we are not only social but also *spiritual* beings. Spirituality is not some people are born with and others not. It is an intrinsic attribute of the human being.

The problem is that although we are all spiritual in essence, only few of us are truly spiritual in action. Our spirituality needs to be deepened so it would be effective. Gandhi, who advised us to be the change we want to see, was a deeply spiritual person himself. He knew what he was talking about. If you want to be the change you wish to see in the world, you have to be a spiritual person yourself.

To become a more spiritual person, we need to unblock our intrinsic spirituality—free it from the layers of prejudice that accompanies this way of thinking and being. Spirituality is viewed by today's matter-of-fact individuals as a form of idealistic adventuring. Spiritual tenets are but imagination; wishful thinking at best.

Our native spirituality needs to be unblocked; we must not dismiss spiritual insights with a wave of the hands. Spirituality has been around for thousands of years, and the insights it conveys have been examined and tested by dozens of otherwise different cultures. These insights are needed to counterbalance the narrow pragmatism that dominates today's world. And spirituality is needed to access fresh insights, effective responses to the challenges of our crisis-bound world. Spiritual insights affirm our own deep origins, and our connections to one another and the universe. These are much needed insights in a rudderless world.

How do you become an authentic spiritual person yourself? The self-taught spiritual master Pierre Pradervand suggests that, to begin with, you ask yourself two questions:

What is my true priority in life, what am I really seeking?
Finding personal serenity? Helping to build a win-win world that works for all? Preparing the next reincarnation? Getting ahead in life, socially and materially?

What is the real motivation of my spiritual search?
Finding inner peace? Expressing unconditional love? A life of service? The elevation of the collective level of consciousness? Or just my own enlightenment?

Fostering and developing your spirituality is one of the most important and practical things you could do. We will not solve the problems of poverty, joblessness, ecological unsustainability, global warming, terrorism, and violence—to mention but a few—unless and until there is a critical mass of people who rise above the superficial pragmatism that dominates today's world. These and other problems are symptoms of a curable disease. The disease of a flawed mindset. We can heal ourselves by developing a deeper spirituality.

We need to reconnect with nature, life, and the universe. This connection is important because the norms for health are accessible through deep spirituality. These norms are not perceived by the rational mind: they are screened by the superficial pragmatism that dominates the modern world. A deeper spirituality overcomes this filter and allows the norms of health and oneness to appear in our consciousness.

Repeated experiments show that the kind of information that discloses the norms of health is not the ordinary kind of information, and it does not appear at the ordinary frequencies of the EEG (electroencephalographic) spectrum. Nonordinary information surfaces to waking consciousness well below the frequency range of sense-perceivable reality: in the Alpha domain, and even below, in the Theta and Delta range. Spiritual persons are more likely to resonate with this nonordinary but essential information than rationally disposed individuals. The latter seldom accede to information below the Beta level.

The norms for wholeness and health are masked in the modern world: they are overlaid by a unilateral focus on our everyday commerce with the world around us. We need to go deeper if we are to rediscover them in our consciousness. We need a deep, effective form of spirituality. This is the way to healing, health, and normalcy, because the problems of today's world are largely due to the masking of the nonordinary, but extraordinarily important, information available to us at the deep regions of the EEG spectrum. This information deserves to be recognized and brought to your waking consciousness; it may contain important guidance for your life.*

The conclusion is evident: there is a real and urgent need for spirituality in today's world. If you aspire to be the change you wish to see in the world, you must upshift your spirituality. This is an essential condition of becoming the person whose "be-ing" could change the world.

---

* An important discovery should be noted in this connection. Scientists at the HeartMath Institute found that the human heart has its own powers of perception. It picks up and processes information typically in the low frequency Delta domain. In that range of frequency, information is not usually available through the sensory channels, as people whose brain functions are in this frequency range of are normally in deep sleep. However, some of this Delta-based information seems transmitted from the heart to the brain through the process known as heart–brain coherence. When there is such a highly-tuned connection between the heart and the brain, information from the heart reaches waking consciousness. This suggests that the perceptions achieved by deeply spiritual people originate with their heart. If so, opening your heart is not just a poetic metaphor: it describes a real process of perception, a process with an extended range. It could be a major and eminently practical factor in your becoming that "*be*-ing" who changes the world.

CHAPTER 6

UPSHIFT YOUR SCIENCE

We now turn to science. The worldview coming to light in cutting-edge science is not, as widely suspected, a radically different, and perhaps even contradictory, in regard to the view maintained by the spiritual traditions. On the contrary, many of the tenets of quantum physics and quantum cosmology reaffirm age-old spiritual beliefs, such as, among others, the existence of space – and time-transcending interconnections, and the evolutionary drive toward coherent states of existence.

Upshifting your spirituality and upshifting your science are twin pillars of the worldview that substantiates and empowers today's attempts to upshift your world.

There are two ways to live your life: as if everything is a miracle, or as if nothing is.
Albert Einstein

The Miraculous Universe

Reviewing the insights science offers today regarding the fundamental nature of the world yields a surprising conclusion. The real world is not what we thought it was. It is a miraculous world of universal interconnection and embracing coherence.

We now review in turn the emerging insights of science regarding the true nature of the physical world, the nature of the living world, and the nature of consciousness.

The Nature of the Physical World

Classical physics gave us a mechanistic and atomistic view of the world. It reposed on Newton's universal laws of nature, as stated in his *Philosophiae Naturalis Principia Mathematica* in 1687. These laws became the foundation of the worldview of the modern age. They demonstrate with mathematical precision that material bodies are made up of mass points, and that they move according to mathematically expressible rules on earth, while planets rotate in accordance with Kepler's laws in the heavens. The motion of all masses is determined by the conditions under which motion is initiated, just as the motion of a pendulum is determined by its length and its initial displacement, and the motion of a projectile is determined by its launch angle and acceleration.

But classical physics is not the physics of our day. Although Newtonian laws apply to objects moving at modest speeds on the surface of the earth, the conceptual framework by which these motions and other observed phenomena are explained has radically changed. Today the smallest measurable and distinguishable units of the physical universe are not material entities, but quantized vibrations in a universal field. These are the quantum particles, or quantum waves—because they have both a corpuscular and a wave-aspect. The quanta themselves are made up of unobservable but theoretically distinguishable units called quarks. Quarks and quanta are intrinsically and instantly interconnected throughout space and time.

The idea of instant and intrinsic interconnection originated with the concept of entanglement advanced by Erwin Schrödinger in the 1930s. A seemingly metaphysical idea, its physical reality has been demonstrated over and over again in controlled experiments. Physicists have accepted the strange fact that quarks and quanta, and the structures built of them, are intrinsically entangled with one another. In its totality, the physical universe is an intrinsically and instantaneously interconnected whole—a view very different from the Newtonian universe of separate mass points.

The mechanistic–atomistic view of the universe was not the view held by Newton himself. Current research unearthed important studies where Sir Isaac Newton expounds a numerological, astrological, and inherently spiritual concept of reality. It appears that for Newton himself, researching the spiritual aspects of the world was more important than researching the physical aspects. This has been forgotten in the heat of the enthusiasm with which Newton's contemporaries greeted the scientific materialistic concept he developed, with its mathematically expounded laws and relationships.

Newton's followers inflated Newton's mathematical–mechanistic theory into an entire worldview. This paid off in the practical realm: the first industrial revolution has been squarely based on the root concepts and equations of Newtonian physics, testifying to their correctness. That they did not testify to their inflation into a general view of reality has not been immediately evident.

However, the inflated Newtonian worldview (which, as we now realize, was not Newton's own worldview but that of his followers) began to crumble at the end of the nineteenth century. The supposedly indivisible atom proved fissionable to a bewildering variety of components that, in the subsequent decades, dissolved in swirls of energy. Max Planck discovered that light, like all energy, is quantized and is not a seamless stream. Faraday and Maxwell came up with the theory of nonmaterial electromagnetic fields, and Einstein postulated that all events in space and time can be integrated in a four-dimensional continuum called spacetime.

The death knell of the Newtonian worldview was sounded in the 1920s with the advent of quantum physics. The quanta of light and energy that surfaced in ever more sophisticated experiments did not conform to our expectations of the behavior of macroscale objects. Their behavior proved to be more and more weird. Einstein, who received the Nobel physics prize for his work on the photoelectric effect (where streams of light quanta are generated on irradiated plates), did not suspect, and was never ready to accept, the weirdness of the quantum world. But physicists investigating the behavior of these packets of light and energy found that, until registered by a detecting instrument or another act of observation, quanta have no specific position, nor do they occupy a unique state. This seemingly weird proposition had to be accepted: the basic units of physical reality have no uniquely determinable location, and exist simultaneously in a superposition of several potential states.

Unlike the mass points of Newtonian physics—which are unambiguously definable in terms of force, position, and motion—the definition of the state of quanta had to be given by a wave function that encodes the superposition of all the states the quantum can potentially occupy. A quantum of light or energy (and subsequently, also of force) proved to be indeterminate as to the choice between its potential states. It manifests properties either as a wave or as a particle, but not as both. And its properties cannot be measured at the same time: if we measure position, for example, energy becomes blurred; and if we measure energy, position becomes indistinct.

However, as soon as it is observed, the quantum's indeterminate state is dissolved: it becomes determinate. It "actualizes" one of its potential states. In the language of physics, the quantum's superposed wave function collapses into the wave function of a classical particle.

What the weirdness of the quantum means in terms of our understanding of the nature of physical reality has been debated for nearly a century. The main points were made by pioneering physicists such as Niels Bohr, Werner Heisenberg, Louis de Broglie, and Erwin Schrödinger. Bohr advanced the principle of complementarity: a quantum has not one but two complementary aspects: wave and particle. Whether it appears as a wave or as a particle depends on the kind of questions we ask and the kind of observations we make. Heisenberg in turn put forward the "principle of uncertainty" according to which at any given time only one aspect of the quantum is measurable; a complete description is forbidden by nature.

The physical origins of complementarity and the interdiction of complete observation remained mysterious. According to Bohr, the very question whether the quantum is a wave or a particle "in itself" is not meaningful and should not even be asked. Quantum physicists must accept an intrinsic limitation: they can only deal with, in Nobel physicist Eugene Wigner's telling phrase, *observations*, and not with *observables*.

The Nature of the Living World

For most of the twentieth century, biology, the science of the living world, emulated physics in aspiring to be empirical and precise. In the embracing Newtonian perspective, it ended up being materialistic and mechanistic. The holistic concepts that dominated the biology of the nineteenth century were condemned as speculative and "metaphysical." The mainstream biology of the twentieth century claimed that life emerges as the result of a random process without inherent aim. Chance-based alterations in the genetic structure of species are exposed to natural selection and generate the forms of life we encounter.

In the second half of the twentieth century, pioneering biologists attempted to transcend these notions. They began to consider the organism as a complex system with its own dynamics and guidance system. The organism is to be considered a whole system made up of interacting parts, such as cells, organs, and organ systems. These parts can be analyzed individually, and the analysis can show how their interaction produces the functions and manifestations of the living organism.

The above conception gave rise to molecular biology and modern genetics and encouraged the trend toward genetic engineering. The initial success of these methods and associated technologies was considered evidence for the soundness of the concept.

However, in the late twentieth century further developments took place. Leading biologists noted that the alternative to mechanism is not a return to the nineteenth century notions of vitalism and teleology, but the development of an organismic approach to the phenomena of life. Their approach has been adopted by the leading process thinkers, among them Henri Bergson, Samuel Alexander, Lloyd Morgan, and Alfred North Whitehead as an embracing philosophy of nature. Whitehead's concept of the organism—the "actual entity"—as a fundamental metaphor for entities in both the physical and the living world served as the rallying point for the post-Darwinian schools of the new biology.

The organismic approach maintained that organisms have a level and form of integrity that cannot be fully understood by studying its parts and the interaction of its parts. The classical holistic concept, "the whole is more than the sum of its parts," was resuscitated. It became evident that when the parts of the organism are integrated within the whole organism, properties emerge and processes take place that are not the simple sum of the properties of the parts. The organism cannot be reduced to the interaction of its parts without losing these "emergent properties." They are the very features that make the biological organism a living entity.

"Coherence" is the concept that best expresses the new holism in biology. An organically coherent living system is not decomposable to its component parts and levels of organization. In the words of biophysicist Mae Wan Ho, such a system is dynamic and fluid, its myriad activities self-motivated, self-organizing, and spontaneous, engaging all levels simultaneously from the microscopic and molecular to the macroscopic. There are no controlling parts or levels, and no parts or levels to be controlled. The key concept is not control, but *communication*. Thanks to the constant communication of the parts in the organism, adjustments, responses, and changes needed for the maintenance of the whole can propagate in the organism in all directions at once.

Similarly to the entanglement of quanta in the physical world, in the living organism instant correlations enable changes that propagate throughout the system, making even distant sites neighboring. This finding is incompatible with the mechanistic concept of the organism, where the parts are separate from one another and have definite boundaries and simple location in space and time.

Coherence in the living realm appears to be universal. It ranges from the smallest element in the organism to the full range of life on the planet. It encompasses multi-enzyme complexes inside cells, the organization of cells in tissues and organs, the polymorphism of living species within ecological communities, and the entire web of life in the biosphere. It ensures the coordination of the biosphere's myriad organic and ecological systems, and their coherent coevolution.

The concept of coherence in the living realm conflicts with the mechanistic assumption of chance interactions among independent elements. The new concept is more than a philosophical or metaphysical tenet: there is evidence that pure chance (which would require the complete absence of causal links between the organism and its surroundings) is a theoretical construct: it is never the case in the real world.

The evidence on this score is wide-ranging. Random mutations in the gene pool cannot explain the evolution of life even in its earliest epochs—complex structures have appeared on earth within astonishingly brief periods of time. The oldest rocks date from about 4 billion years, and the earliest and already highly complex forms of life (blue-green algae and bacteria) are more than 3.5 billion years old. The classical theory cannot tell us how this level of complexity could have emerged within the relatively short period of about 500 million years. A chance mixing of the molecular soup in the primeval earth would have taken incomparably longer to produce the phenomena we observe. (The assembly even of a primitive self-replicating prokaryote—a primitive, non-nucleated cell—involves building a double helix of DNA consisting of some one hundred thousand nucleotides, with each nucleotide containing an exact arrangement of thirty to fifty atoms, together with a bilayer skin and the proteins that enable the cell to take in and process food. This structure requires an entire series of coordinated reactions finely tuned with one another. Producing it is unlikely to be the result of chance interactions among separate elements. Random mutations and natural selection may account for variations within a given species, but not for the evolution of complex living systems in the given finite time frame. Mathematical physicist Fred Hoyle pointed out that the probability that evolution would occur by chance is as likely as a hurricane blowing through a scrap yard assembling a working airplane.)

Life comes about by massive and highly coordinated innovations in the genome, rather than by chance-based piecemeal variations in the genetic code. Genes do not work in isolation: the function of each gene is dependent on the context provided by all the others. The whole "ecology of genes" exhibits layers and layers of feedback regulation, originating both with the physiology of the organism and in the relationship of the organism to its environment. These "epigenetic" regulations can change the function of the genes, rearrange them, make them move around, and even mutate them. Major mutations are not due to a haphazard recombination of genes; they are responses of the epigenetic network of the organism to the chemical, climatic, and other changes generations of living organisms have experienced in their milieu. (The emerging insight combines a long-discredited thesis of Jean Baptiste Lamarck [that the changes experienced by organisms can be inherited] with a pillar of the theory of Darwin [that inheritance is mediated by the genetic system]. It now appears that that the experiences of the organism in its milieu affect subsequent generations of organisms. But this is not because the experiences are directly communicated from one generation to the next, but because they leave their mark on the epigenetic system. Through the modification of the epigenetic system, the mark of the experiences is handed down from generation to generation.)

The discovery of subtle links between the genome and the organism, and between the organism and the world around it, suggests that the living world is not the harsh domain of classical Darwinism, where every species competes for advantage with every other. Rather, life evolves through what biologist Brian Goodwin called "the sacred dance" of the organism with its milieu. Subtle strains of this dance extend to all the species and ecologies in the biosphere.

The findings of contemporary biology tell us that we are not Newtonian machines. We are not separate from each other and from our environment; we are part of an interconnected system with an intrinsic evolutionary drive and space – and time-transcending interconnection. We are elements in a quasi-living self-evolving universe, where every element interacts with every other, and jointly creates systems of great and ever-increasing complexity and coherence.

The Nature of Consciousness

It is through our consciousness that we experience the world. We never experience the world directly, except (perhaps) through intuition or enlightenment. Information about the world comes to us through the flow of sensations that accompanies us throughout our lives—the flow we experience as our consciousness.

Can we put the question regarding the nature of consciousness on a par with the question concerning the nature of the real world—the physical world and the living world? The answer is that we can. We have good reason to believe that consciousness is a real phenomenon in the real world, taking its place among energy, frequency, and information. In the final count, the reality of consciousness is better substantiated in natural science than the reality of matter. The latter, as we have seen, is seriously questioned.

Consciousness is at the same time the most familiar and the most mysterious element of our lives. Consciousness is mysterious because it is not clear what it is and where it comes from. Is the flow of sensations that makes up our consciousness generated in, and confined to, our brains? Or does it extend in some way beyond our bodies and brains?

Until a few years ago, nobody other than deeply spiritual or religious people would have subscribed to the proposition that consciousness is more than a product of the workings of the brain.

But today there is an insight dawning among leading thinkers and scientists—the insight that consciousness is nonlocal. It exists in association with the brain, but it is not produced by and confined to the brain.

The classical concept of consciousness

In mainstream science, the accepted concept of consciousness is the classical concept, consistent with the physics of Newton and followers. In their view, there is no place for consciousness in the universe: in the last count, all that furnishes the universe is bits of matter moving in space and time. Consciousness is an epiphenomenon: something generated by a real phenomenon but not real in itself. In this respect, consciousness is like the electricity generated by a stream of electrons in a turbine. The electrons are real, the turbine is real, but the electricity generated by them is a secondary phenomenon. It disappears, after all, when the electrons cease to stream in the turbine. The existence of electricity is contingent on the working of the turbine, and the existence of consciousness is contingent on the working of the brain. Consciousness no more exists in a dead brain than electric charge exists in a stationary turbine.

We do not see, hear, or taste electricity; we know it only by the effect it produces. This is said to be the same with consciousness. We experience the flow of sensations, feelings, and intuitions we call consciousness, but we do not perceive consciousness itself. No amount of scrutiny of the brain will disclose anything we could call consciousness. We only find gray matter with networks of neurons firing in sequence. In creating the sequence of the flow of electrons the brain generates the sensations we experience: feelings, and volitions. When the brain's operations are damaged or reduced, consciousness is distorted, and when the brain stops working, consciousness ceases.

For the classical concept there is nothing miraculous, or even mysterious, about the presence of consciousness in the universe. Human consciousness is the product of the workings of the human brain.

The new concept of consciousness

The turbine concept of consciousness is a hypothesis and, as other hypotheses, it can be upheld if the predictions generated by it are confirmed by observations. In this instance, the generally cited prediction is that when the brain stops working, consciousness vanishes.

On first sight, this prediction is confirmed by evidence: observations indicate that when cerebral functions come to a halt, consciousness disappears. This, of course, is not directly experienced, but it is a reasonable inference from what we do experience. People who are brain-dead do not seem to possess a working consciousness.

The above claim does not admit of exceptions. We can no more account for the presence of consciousness in a dead brain than we could account for the presence of electric charge in a stationary turbine. Evidence to the contrary would place in question the basic tenet of the classical concept of consciousness. But evidence to the contrary does exist. It surfaced in clinically controlled, rigorously protocolled experiments. There is real and credible evidence that in some cases consciousness does not cease when brain functions do.

The most widely known form of the evidence is furnished by people who have reached the portals of death but returned to the ranks of the living. In some cases their consciousness appears to persist even when their brain functions are "flat." Many temporarily brain-dead people report having had conscious experiences during their near-death episode. NDEs—near-death experiences— are surprisingly widespread: in some cases they are reported by up to 25 percent of the people who experienced a condition near death.

Are NDEs cases of veridical recall, or are they fantasy? Repeated experiments suggest that they are veridical. NDE reports have been confronted with experiences the subjects would have had if their brain would have been functioning normally, and in a significant number of cases the recalled experiences and the "would have" experiences exhibit a remarkable and more than random match.

There are indications that conscious experience persists not only during the temporary cessation of brain function, but also in its permanent absence: when the individual is fully and irreversibly dead. These surprising experiences became known as ADEs: after-death experiences. The evidence for them is offered by mediums in deeply altered states of consciousness. In these trance-states they appear able to communicate with deceased persons. They "hear" the deceased people recount their experiences after they died.

Reports of ADEs have been subjected to systematic scrutiny, exploring the possibility that the mediums would have invented the messages, or picked them up from living persons through some form of extrasensory perception. In a non-negligible number of cases, the theory that they were invented by the mediums could be ruled out: the messages contained information the mediums were unlikely to have invented themselves. And accessing such information through ESP (extra-sensory perception) is a logical assumption, but ESP itself is in need of corroboration by reliable evidence.

Given the mounting evidence, we are logically obliged to accept that some near-death and after-death experiences are authentic. This calls for a radical revision of the mainstream concept of consciousness.

The new concept is that consciousness is more than a product or by-product of brain function. "Our" consciousness is a local and temporary manifestation of a consciousness that is a real element in the real world. More and more consciousness researchers join pioneering brain scientists, psychologists, and psychiatrists who do not hesitate to make this assertion. Consciousness exists beyond the brain.*

The beyond-the-brain concept of consciousness has been maintained by a number of world-renowned scientists. Erwin Schrödinger, for example, did not hesitate to say that consciousness

---

* For example, Pim van Lommel *Consciousness Beyond Life*. Harper Collins, 2010, and Edward Kelly, Emily Kelly, Adam Crabtree, Alan Gauld, Michael Grosso, Bruce Greyson, *Irreducible Mind: Toward a Psychology of the Future*. Lanham, Rowman and Littlefield, 2007, 2010.

does not exist in the plural: the overall number of minds in the world is one. In his last years, Carl Jung came to a similar conclusion. The psyche is not a product of the brain and is not located within the skull; it is part of the one-universe: of the *unus mundus*. In David Bohm's quantum cosmology, the roots of consciousness are traced to the deeper reality of the cosmos: the implicate order. A number of contemporary scientists, such as Henry Stapp, maintain and elaborate this concept. Consciousness is nonlocal: it is a cosmic phenomenon.

In the emerging concept, the flow of sensations we call consciousness is a manifestation of the cosmos—an element that is as real as energy, frequency, amplitude, phase, and information, and more real than matter. The brain is not a material turbine that generates consciousness, and consciousness is not its product or by-product. Consciousness is a cosmic phenomenon. The brain is only its receiver.

The world we perceive is a local manifestation of cosmic consciousness. As perceptive scientists and philosophers have long told us, the cosmos not only *has* consciousness, but *is* consciousness.

The ancient insight has been brilliantly stated by the modern-day yogi Swami Vivekananda. "We now see that all the various forms of cosmic energy, such as matter, thought, force, intelligence, and so forth, are simply the manifestations of that cosmic intelligence, or, as we shall call it henceforth, the Supreme Lord. Everything that you see, feel, or hear, the whole universe, is His creation, or to be a little more accurate, is His projection; or to be still more accurate, is the Lord Himself."*

Erwin Schrödinger, and numerous quantum physicists and consciousness researchers, said that there is no sense in which we can speak of consciousness in the plural. The cosmos is one, and if consciousness and the cosmos are the same, then consciousness is one.

---

* Swami Vivekananda, *Complete Works* Vol 2, p: 462

We reach a basic conclusion. Consciousness is cosmic, because consciousness is not *in* the cosmos. Consciousness *is* the cosmos. Which is to say that the cosmos *is* consciousness.

A note on the difficulty of expressing this proposition

The Western subject-predicate language is not adapted to expressing the proposition that cosmos and consciousness are one and the same. The ordinary mode of stating the relationship between them is to claim that the cosmos "has" (possesses or includes) consciousness. If so, we have both a subject and a predicate, yet in the real world there is no subject apart from the predicate. In the new science context, we can speak of *hardness* and *blueness*, but not of hard and blue "things," The existence of "things" in the universal quantum field is purely speculative.

CHAPTER 7
UPSHIFT YOUR WORLD

Let us assume that you have taken the first steps on the way forward (per chapter 4); have upshifted your spirituality (per chapter 5); and have also upshifted your science (per chapter 6). Then you are ready to tackle the biggest upshift of all: the upshift of the world in which we now find ourselves.

The upshift of our world needs to be carefully planned and prepared. It will be useful first to build a prototype—prepare a Plan minus-A of what we seek to accomplish. This doesn't mean doing science fiction; it means drawing the contours of a world that is both desirable and achievable.

Mapping out such a world is the objective of this chapter. This is a long-term project—a desirable and attainable world is not likely to come online this year or the next. But it could be created, at least in basic terms, by the year 2050, and even by 2030. Is this overly optimistic? Time will tell. We know that a global crisis is an effective taskmaster: changes take place fast when the urgency is high. As Dr. Johnson said, nothing concentrates the mind so wonderfully than the knowledge that you will be hanged in the morning. We will not be hanged tomorrow morning, the persistence of *homo sapiens* will not be cut short so soon. But the very existence of the less privileged members of the species could be seriously in question in a decade or less. We need to act fast.

Let us imagine what a better world could look like in a positive but not utopian timeframe. Then we could calibrate the measures we need to take to achieve that world in practice.

Responsible Life on the Planet: An Achievable Vision

Dateline 2030

Social and Political Organization

The world in 2030 is globally whole and locally diverse. Sovereign nation-states, the inheritance of the modern age, have given way to a transnational world where nations are only one, even if an important, level of political organization, without claims to sovereignty.

The 2030 world is networked, but not monolithic. It is organized as a Chinese box of administrative and decision-making forums, where each forum is embedded in each higher forum but has its own sphere of authority and responsibility. The political world is not a hierarchy, for the decision-making forums at the various levels have their own autonomy and are not subordinated to higher levels.

In some areas—including trade and finance, information and communication, peace and security, and environmental protection—decision-making is entrusted to global forums. This, however, allows a significant level of autonomy on local, national, and regional levels. Taken in its totality, the political world is a "heterarchy": a multilevel sequentially integrated structure of distributed decision-making aimed at global cooperation combined with regional, national, and local spheres of authority and leadership.

The diverse yet cooperative world is a sequence of self-reliant communities with multiple links of communication and cooperation. Individuals join together to shape and develop their local community. These communities participate in a wider network of cooperation that includes, but does not cease at, the level of national states. Nation-states are themselves part of regional social and economic communities, coming together in the United Regions Organization, the global-level body created through the reform of the United Nations Organization. Its members are not nation-states but the continental and subcontinental economic and social unions that integrate the interests and programs of nation-states. These include the European Union, the North American Union, the Latin American Union, the North-African Middle-Eastern Union, the Sub-Saharan African Union, the Central Asian Union, the South and Southeast Asian Union, and the Australian-Asia-Pacific Union.

The principle of subsidiarity holds sway throughout the multi-level system: decisions are made on the lowest level at which they are effective.

The *global level* is the lowest level in regard to ensuring peace and security and regulating the global flow of goods, money, and knowledge. It is also the level for coordinating the information that flows on global networks of communication. Its objective is to harmonize policies dedicated to ensuring the integrity of the processes that maintain equilibrium in the biosphere.

The *regional level* is indicated for the forum for policies that coordinate the social and political aspirations of nations. Regional economic and social organizations provide the forum for the representatives of member nations to coordinate their interests and aspirations in view of resolving the problems of their people.

The *national level* is appropriate for the local tasks and functions traditionally entrusted to national governments. National organizations operate without claiming unconditional sovereignty for themselves; they are embedded in the regional and global-level forums and take due account of the decision made by them.

On the *local level*, forums bring together the elected representatives of urban as well as rural communities. They coordinate the workings of the social and political institutions of towns, villages, and rural regions within the framework of the administrative and decision-making system composed of forums on the national, the regional, and the global levels.

Lifestyles

Some people are well off, but nobody is superrich. Simpler lifestyles are the rule and not the exception; they are the fruit of an upshifted culture that guides people's aspiration for living a healthy life with consideration for others and without ostentation.

The diversity of lifestyles finds expression in the sphere of interpersonal communication, in contact with nature, and (in individual cases) in contact with a consciousness or reality. The basic aspiration is personal growth and development in the embrace of concentric spheres of life and decision-making starting with the family and extending sequentially to local community, region, and nation, to the global community of all nations and regions.

People live longer and healthier but do not trigger an explosion of the population. They realize that it is irresponsible to create families larger than the replacement level. In most parts of the world, two-child families is the popular dimension of fertility.

The human population is moving toward equilibrium between fertility and mortality at a low and for the present sustainable level. This offers benefits to everyone. With modestly sized families, parents are better able to care for their children and ensure that they grow into healthy persons with sufficient education and access to information to live well.

Lifeways are becoming ecologically conscious. As people are reoriented from self-centered satisfactions brought by individual consumption toward personal growth and interpersonal development, energy and material requirements in the world are diminishing to more sustainable levels. And, as people work together to improve their shared living and working environment, community life enjoys a renaissance. People, both women and men and young and old, rediscover a deeper dimension in their consciousness, the dimension of the evolutionary impetus. They search for coherence and oneness, and ultimately love, as the grounding values of their existence.

Morality

Lifeways remain socially, culturally, and geographically diverse. Religious beliefs, cultural heritage, technological development, levels of industrialization, climate, and nature are all factors that enter into and enrich the panoply of lifeways. Yet, beyond the diversity there is a shared morality. Personal and interpersonal growth and development are to respect the limits of an acceptable quality of life for all people in the human community.

The shared morality goes beyond the pragmatic liberalism proclaimed by the Anglo-Saxon philosophers, among them Jeremy Bentham, John Locke, and David Hume. In their view, people can pursue their interests as long as they respect the rules that safeguard life in a civilized society. "Live and let live" is the motto. People can live in any way they please, as long they do not break the law.

In the year 2030, people realize that in a highly interdependent and delicately balanced world, the classical forms of liberalism are misplaced. Letting everyone live as they please as long as they keep within the law entails a serious risk. The rich and the powerful may consume an ever more disproportionate share of the resources to which the poor also have claim, and may harm the environment that is a common resource for all.

Rather than "live and let live," a code of behavior is coming to light that is better adapted to conditions on the planet than classical liberalism. It replaces liberalism's "live and let live" with Gandhi's "live more simply so others could simply live."

Gandhi's injunction is further specified. The principal concern is not with the intrinsic simplicity of the way people live, but with the impact of the way they live on others, and on nature. This impact must not exceed the capacity of the planet to provide for the needs of all people. Simple lifestyles are to be favored not just because of an intrinsic preference for simplicity, but because simple lifestyles are more likely to remain within the limits of adequate human resource availability in the biosphere. The new moral code takes into account the impact of the human species on earth—a planet of great wealth but of finite resources. The emerging moral code is summed up in the injunction: *live within the limits of the resource-use that permits others to live as well.* As the new moral code takes hold in the mainstream of society, cutthroat competition for selfish and self-centered ends is replaced by a spirited rivalry in the context of shared goals and common interests.

Expecting that people abide by the new moral code does not call for people to become selfless angels. Living in a way that permits also others to live does not mean that everybody could live in the same way with the same material standard of living. There are differences in wealth and standards of living, but these differences are moderated by the moral injunction regarding the limits of how we can live our lives.

The new morality respects the right of all people to live a life of dignity, spared the deprivations that plagued the poor in the recent past. Being moral does not call for being self-denying—it allows everyone to strive for excellence, for beauty, and personal growth. But in the context of an interdependent and finite planet, the enjoyments and achievements of life are to be defined in relation to the quality of enjoyment and level of satisfaction they provide, and not in terms of their monetary cost and the quantity of materials and energy they consume.

Beliefs

The lessons of the faulty beliefs that hallmarked the world of the early twenty-first century have not been forgotten: people know that what they believe shapes the way they think and act, and can create a faulty world as well as a better one. A periodic reexamination of our core beliefs is a precondition of safeguarding the better world we seek to create.

The beliefs and convictions that emerge in the current reexaminations are very different from the beliefs that hallmarked the past world. The decisive edge of the beliefs that shape the world we seek can be summed up in a handful of bullet points. This, then, is the essence of our updated beliefs:

- We are part of the web of life on earth, and the web of life is part of us. We are what we are in our communication and communion with the beings that emerge and evolve in the web of life of this planet.
- We are more than a skin-and-bone material organisms. Our bodies with their cells and organs are manifestations of what is truly us: self-sustaining, self-evolving, dynamic beings arising and evolving in interaction with all other self-evolving dynamic beings around us.
- We are one of the highest, most evolved manifestations of the drive toward coherence in the universe. Our core essence is this universal drive. Recognizing it and aligning with it is both our duty and our privilege as conscious beings.
- There are no absolute boundaries and divisions in the world, only phase transitions where one set of relations yields prevalence to another. In our bodies, the relations that integrate our cells and organs into a dynamic and coherent whole are prevalent. Beyond our bodies, the relations that drive toward coherence and wholeness with the communities of living beings gain prevalence.
- The separate identity we attach to people is a convenient convention that facilitates interaction with them. The whole gamut of concepts and ideas that separates our identities from the identities of other persons is but an arbitrary convention. There are only gradients distinguishing individuals from each other and from their environment, and no categorical divisions and boundaries.
- In the final count, there are no "others" in the world: we are all living beings and we are all part of each other. Our family and community are just as much who we are as the cells and organs of our bodies.

- Collaboration, not competition, is the royal road to sustain us and all the beings who inhabit this planet. Collaboration calls for empathy and solidarity, and ultimately for love. Collaboration inspired by love is the way to achieve health and well-being for ourselves, and for all the beings with whom we share the planet.
- Attempting to advantage the beings we know as "us" through ruthless competition with the beings we know as "others" is a grave mistake: it damages the integrity of the embracing whole that frames our lives. When we harm "others," we harm ourselves.
- The idea of advantaging ourselves, even our families, communties, and nations, without regard for the beings we used to regard as "strangers" or "foreigners," needs to be rethought. Patriotism, if it aims to eliminate adversaries by force, and heroism, even in the well-meaning execution of patriotic aims, are unwise and dangerous aspirations. A patriot and a hero who brandishes a sword or a gun is an enemy to everyone, himself or herself included. Comprehension, conciliation, and forgiveness are the hallmarks of courage, not championing the ambitions of some without concern for others.
- "The good" for anybody in the human community is not the possession of riches. Wealth in any material form is but a means for curating our existence in the embrace of the community of life. As exclusively "mine," wealth commandeers part of the resources all living beings need to share. Exclusive wealth is a threat to the communities of life, including the communities of those who hold it.
- The true measure of our accomplishment is the measure of our sharing, not of our having. Sharing enhances existence in the communities of life, whereas possessing creates division, invites competition, and fuels envy. The share society is the norm for the communities of life; the have society was typical only of modern society, and it was an aberration.

- We and our fathers and forefathers have been guilty of the aberration of the human family, and as conscious and responsible members of that family, we recognize that correcting these aberrations is our ineluctable duty. Beyond the pursuit of love and meaning, it is the highest objective we can adopt in our lives.

CHAPTER 8
A THOUGHT TO REMEMBER

D o not forget who you are and what your mission is in life. You are a member of a remarkably evolved species living in the embrace of a remarkably evolved web of life in a remarkably life – and spirit-friendly region of the universe. Your fathers and forefathers have taken a wrong turn in the path of humanity's evolution, and it is your duty to help today's generations to regain the right path. Your mission in life is to further and facilitate this epochal project. Follow your deepest intuitions. *Perceive nature's intrinsic penchant for wholeness and coherence, and let the unconditional love it suggests radiate from your heart and illuminate your mind. Let it inspire how you live, and how you relate to others and to the planet.*

You and I, we are the movers and the shakers, the music makers. We are the power behind the upshift to a world where all people can live and thrive, and where all can lend their voice to the supreme symphony of life on earth.

> We are the music makers,
> and we are the dreamers of dreams,
> Wandering by lone seabreakers,
> And sitting by desolate streams;
> World-losers and world-forsakers,
> On whom the pale moon gleams:
> *Yet we are the movers and shakers*
> *Of the world for ever, it seems...*

We, in the ages lying
In the buried past of the earth,
Built Ninevah with our sighing,
And Babel itself in our mirth;
And o'erthrew them with prophesying
To the old of the new world's worth;
For each age is a dream that is dying,
Or one that is coming to birth.

Arthur O'Shaughnessy

ABOUT THE AUTHOR ... IN
HIS OWN WORDS

**Ervin Laszlo interviewed by David Lorimer
for the Scientific and Medical Network
and the Institute of Noetic Sciences***

David Lorimer

I will start by asking you about your childhood and early background. How you became a concert pianist, and about your mother, who was a pianist herself. Can you tell us something about the early phase of your life? How did you decide to become a pianist?

Ervin Laszlo

Actually, nobody asked me—I wasn't consulted about what I wanted to be. I grew up from the time I was four or five years old with the idea that I'm a pianist. I never thought I would be anything else. Nobody around me thought I would be anything else.

It was not that an uncalled-for arbitrary design would have pushed me into something that was not really good for me. It turned out that I have a feeling, a real and unusual talent for music.

I could live and not only hear music. I could see how a piece develops. I could hardly read musical scores since I very rarely consulted them. In my childhood my mother would play a piece

---

* Reprinted by kind permission of the Scientific and Medical Network (U.K.) and the Institute of Noetic Sciences (U.S.).

107

for me, and then I would play it, and I would continue playing it. It was purely intuitive. Music was something that was given. I could speak through music, just as people speak their own language. I could express myself through music. Music was a way of existence for me.

Music is a language and great musical works have a particular completeness to them. They come to a close and have a perfection to them. Music is a deeply emotional experience based on the feeling of perfection.

I used to practice first to memorize a particular piece and get my fingers to play it. Then Mother would say, now play it with feeling. That's how I would play it, not just playing the notes. Playing what the music tells me.

This was in my childhood, how I started. My mother was a musician and my uncle was a musician as well. He was a violinist. But at the same time, he was a philosopher. Ours was a very musical and intellectual household.

David

Do you have a favorite piece that you love playing more than any other? I know you still have a piano and you probably still play every day.

Ervin

My favorite piece is what I'm feeling myself into at the time. It's not an abstract choice, not something I would like to hear on a CD or by going to a concert. It's what I play when it comes to me. I just sit down and play. That's what my favorite is—just what is emerging for me at the time.

David

So that's how it is. Now tell us the story of your playing the Beethoven's Waldstein Sonata in a concert and meeting the person who then published your writings.

Ervin

That person was a publisher who then published the notes and thoughts I wrote just for myself. That was what brought me to the realization that I could be something else in life than a concert pianist.

David

I just wanted to highlight the musical part of your life because a lot of people are not familiar with it. Now we come to the more academic part of your life. How did that start? I think it was the Waldstein Sonata you were playing on the concert stage when you realized that you weren't sure where you were in playing the second movement, whether you were at the end of the movement or in the middle.

Ervin

Yes, for a moment, I felt completely lost. Frightened. Frozen, because it's such a crucial choice. I was performing in the Beethoven House, in the hall dedicated to Beethoven in Bonn, his native city. I was playing a cycle of Beethoven's piano works. Performing his music in a concert hall dedicated to him was a tremendous responsibility. There I was in the middle of the sonata's beautiful slow movement, and I didn't know whether I'm playing the theme for the first time or the second. Because this particular theme recurs at the end. I just played on, and hoped that I was neither repeating the theme unnecessarily, or miss playing it altogether. If the public's response to it proves to be continued listening, I am all right. Because missing the entire movement is even worse than repeating it.

This was a moment of great dilemma. I took a chance and played the whole movement as if for the first time, and nobody got up, nobody laughed or made any signs. So I happened to get it right. But then, I said, this cannot go on. What alternative do I have in my life to giving concerts?

It so happened that I had just received a message from Professor Schrader, the head of the Philosophy Department at Yale University. He was reading the book I published by that time, titled *Essential*

Society. He was impressed by it and wrote, would you like to come to be a fellow of the Philosophy Department? You would meet interesting people, he said, who could be important for developing your thinking.

This was a great opportunity, but a difficult choice, because it was difficult for me to leave my musical career. I had many concerts booked, entire concert tours planned for many months ahead. How could I just get up and leave all that?

I had to decide, and on the spur of the moment I decided that this invitation came just now was perhaps not purely by chance. Perhaps it was something I was meant to accept and follow up.

I sent off my answer by telegram. It said, I am accepting and coming. (At the time we could only send telegrams to each other for urgent communication, there were no emails and Skype or Zoom messages.) I will be in New Haven at the beginning of the next semester, meaning next September (the incident in Bonn took place in the spring).

I had to cancel my concert booking and start packing. I lived in Switzerland at the time with my wife and young son, and they agreed to stay there during my exploratory adventure. I was to come back and rejoin them and my music career at the end of the semester.

That was the plan. But my coming back was just another interlude. I received several remarkable invitations for academic positions while at Yale, and I accepted the most suitable among them—first a semester teaching aesthetics of music in Bloomington, at Indiana University, and then teaching philosophy at the University of Akron in Ohio. We went back to America together, and an academic life phase started for me. That's the story in a nutshell of my shift from the concert stage to the lecture rostrum.

David

Thank you for sharing this. Now I want to ask you about the philosopher Alfred North Whitehead, because you, like many other people, have been profoundly influenced by his thinking.

Ervin

I was indeed. Whitehead was far ahead of his time, and in some ways people are only just catching up with him. Reading his major book (*Process and Reality*) made a deep impression on me. I picked up other books that referred to him, then went back to Whitehead himself and tried to really get into it. Of his big opus, I read every sentence at least five times. It is very revealing once one understands it. For me, it was saying yes, of course, that's how the world is. It's not just a set of sturdy material, impersonal things, but an organic entity that evolves, develops, and embraces all things. The world for Whitehead is a process of development, of evolution. This has made a tremendous impression on me, I still think in these terms.

Another influence on my thinking was the Belgian Nobel physicist Ilya Prigogine. I met him about twenty-five years ago. He told me that Henri Bergson was a great influence on him in turn. These scientists looked at process and evolution as something fundamental, and also universal.

We now understand that evolution is not limited to genes in living species. Evolution happens throughout the universe. There is nothing in the universe that was the way we find it, everything evolved into what it is now.

David

You speak a great deal about a sense of purpose that would underlie evolution in the world.

Ervin

Yes—I have moments when it's very, very clear to me that not all ways to act and evolve are equally good, and not all are equally right to take. When a course of action is the right one, I have a sense that this is so—this is what I should be doing. Things somehow fit into place. But if things don't fit, sometimes even quite physically refuse to move forward, I feel that it is best to abandon them. Sometimes when I want to jot down a thought, the whole system freezes up. I try it again and again, and if it still doesn't work, something is wrong, it is not the right thing. I rethink that thought, and sometimes the different idea flows on to the keyboard. It feels right, and it works. This is a kind of spontaneous guidance from the universe. I believe in it and I trust it. If I am on the right track, if I do the right thing, this guidance facilitates doing what I intended to do. I just throw the first pebble and see what waves would appear. Sometimes these waves grow and develop and become a lot more than I expected.

David
One of the people you met at Yale was Ludwig von Bertalanffy, the Viennese biologist. Can you give me an idea of your impression of him, and of the influence of his thinking on you?

Ervin
Bertalanffy was the founder of what became known as general systems theory. This term is a source of misunderstanding, because people are talking about general systems as such, as if this were a different kind of system. Actually, the meaning of general systems theory is that it's the general theory of systems. The term "general" applies to the theory, and not to the system. There is no such thing as a general system, only a general theory of systems. The concept "system" applies not only in biology, but wherever evolution takes place. And if most of the entities in the natural world are systems, that means that these entities are not separate and separable, little particles or elements. Evolving things are part of larger wholes, and all these systems, wholes and evolving systems that are parts of wholes, we can regard as systems. The term itself is not important, only what we mean by it. I remember von Bertalanffy saying once, I don't care what they are called, you can call them crêpes suzettes. What matters is that such a system is an integral whole. Its parts stay together and develop and interact with other wholes. Creating more complex and extended "suprasystems."

This was a very avant-garde notion at the time. It is better known today, but it is not really made use of. Leading people in politics and business base their decisions on linear analysis. And if such analysis doesn't penetrate the problems, they call those "wicked problems." They don't realize that complexity in the world calls for systems analysis, otherwise everything turns out to be wicked. In the world of systems, everything is seen to be in evolution, and evolution is toward increasing complexity and coherence. That is not a simple coherence, it's a coherence of interacting parts that are diverse among themselves. When these systems evolve, the problem of understanding them compounds very fast and becomes unmanageable in terms of classical analysis.

But analysis in terms of evolving systems simplifies the situation, because at the level of the highest level system, we don't need to deal with all of the complexity that lies below. We need only to deal with the results of that complexity. A CEO or a prime minister doesn't need to deal with the problem of cleaning the streets and administering exams in the schools. They can deal with the problems of social, economic, and ecological policy selectively. They extract the data that's relevant for their sphere of competence, data that refer to the results of complexity. They only need to deal with that.

Complexity in a system is relative to the position of the perceiver. The more you look at the parts, the more things get complex. If you look at the wholes produced by the interaction of the parts, things become simpler and more manageable.

The overall system is the product of a trend or development that holds sway in the system's environment. If you know the nature of that trend, you can do a more goal-oriented, more relevant analysis.

I'm not worried about dealing with complexity in a system. I try to deal with systems at the level of wholes, recognizing that these wholes are parts of larger wholes. I don't go into the smallest detail—I try to get an overview. Maybe that's a shortcoming. But I don't think so, because at a time of change and transformation, a good overview is essential.

David
Yours is a hugely important role. What you said reminds me that the world's governments and particularly given the way they are related

in the G-9, they act as a rudder. The art of governing is the art of steering, which I think a lot of governments have forgotten.

I want to move on to your time at the United Nations. What sort of reflections do you have regarding the work you did at the UN? What are your thoughts looking back, and especially in the light of where the UN is now and where you think it may be going?

Ervin

Well, the term "United Nations" is practically a contradiction in terms. Because "nation" is defined in the international context as being sovereign, and a sovereign entity is responsible only to itself—its own people and its own territory. How can such a sovereign entity be united? Obviously, if an entity is united with another entity, or group of entities, it cannot be sovereign. Others enter into its decisions and influence its behavior.

Working at the UN, I always opted for promoting the member states being "united" rather than being "sovereign." I tried to point out that every member state has to interact, to join together and not maintain an aloof separateness—the kind of separateness implied by the idea of sovereignty.

This contradiction is a factor in the whole work of the organization. There are the national delegates who are there to safeguard the interests of their own government and state. Then there is the secretariat who tries to create an organization where the states have a sense of unity and work together for the common good. These two things don't mesh—they are at odds with each other. I tried to counteract the nationalism inherent in the representations of the member states. This was natural for me because I have very little national identity myself. I was born in Hungary and left when I was fifteen. I love having a Hungarian identity, because if I don't have a Hungarian identity I cannot have an American, an Italian, or any other identity. These identities need not be mutually exclusive. I have several national identities at the same time, and can also have a global, all-human identity. The latter is the dominant identity in my life.

At the UN, I dedicated my work to bringing about cooperation between the nation-state members through forming a larger unity, a unity between nation-states on the regional level. Interregional cooperation was the program I created and headed for four years at the UN.

I was not a civil servant in the usual sense. I was a special fellow at the Institute for Training and Research, and that meant the freedom to pursue my goals and objectives once they were approved by the secretary general of the United Nations and the executive director of UNITAR. I had an office at the New York headquarters and a budget that allowed me to organize cooperation among research institutes on the regional level.

Then, as I continued work on my projects, I was co-opted into an organizational niche as director at the institute. But I never stayed entirely within my formal role. I was never very good at following rules and I objected to spending 80 percent of my time on political and organizational red tape. I tried to deal with the substance of my projects and I kept out of collateral commitments.

Getting out of bureaucratic obligations was not easy. I didn't play the game of currying favor with the powers that be. I just did what I considered was needed to represent and disseminate the goals and ideals I believed in.

Helping to achieve cooperation among the member states was my goal all along. But it was very difficult to go against the nationalist spirit and be truly global. I would have liked to call the organization "United Peoples" not "United Nations." That was also the vision of Dag Hammarskjöld. He wanted the organization to stand for people, not for states.

David
Let us move on now to the next phase of your life. How did you come to found the Club of Budapest? I understand that this was in cooperation with Aurelio Peccei of the Club of Rome. What were you trying to achieve with this new entity?

Ervin

I was involved with the Club of Rome for many years. I was pleased that its founder and president Aurelio Peccei came to me and suggested that I take an active role in researching and writing a "Report" to the Club. I undertook this, but wanted to reform the Club of Rome at the same time. The Club was highly oriented toward bringing together high-level influential personalities in business and politics. The meetings of the Club were organized to involve prime ministers and presidents and top executives in the corporate world. I thought that this was not the best approach. The Club should be talking to the people and not just leaders. Because wonderful things were said at the meeting about pulling together and cooperation, but come Monday morning, the leaders went back to their country or their offices and did much the same thing they did before. I said that we need to bring together not just business and political leaders, but people whose voices resonate with the people. That means involving artists and writers, sports figures, dancers, musicians, as well as spiritual leaders.

I tried to bring together such people, but with a membership limited to one hundred people, this was not possible within the structure of the Club of Rome. I had to create another entity, a twin Club. This was what I first thought of as an artists' and writers' Club. The goal was to address people's mindsets, their consciousnesses. Artists and writers could do that better than the established leadership.

I called the new mindset "planetary consciousness." This was at the time a very new and daring idea. Most people didn't understand what it means. Does it suggest that the planet has consciousness? This needed clarification. I drafted what I called "The Manifesto on Planetary Consciousness." When I completed the first draft I happened to be in India, serving first as the head of the Auroville Commission in Auroville near Pondicherry, in the southern part of India. The Commission was charged to be the official contact point for Auroville, an independent social and political entity within the territory of India.

The Dalai Lama visited Madras (today Chennai), which was just a few hours' drive away. I asked my friends in Auroville who were being received by the Dalai Lama to introduce me. I went to Madras to be received by His Holiness. I was told that His Holiness has five to seven minutes to offer at the most because he has meetings scheduled all day. I said, let's use that time anyway.

When I sat across from the Dalai Lama he asked me what I had in mind. I said, here is the draft of a manifesto on planetary consciousness. I would like to promote this kind of consciousness in today's world. He said he would like to hear it. I started reading my draft. His secretary was there and took notes. His Holiness kept interrupting, coming up with ideas of his own. We ended up working on it the rest of the day. At the end, we had a basically new draft of the manifesto. That's the draft I took back to Budapest to get the endorsement of the Members of the Club of Budapest.

The first members at that time were the president of Hungary, himself a literary figure of deep spirituality, much like Vaclav Havel, the Czech president at the time. The other initial members were the famous actor playwright Sir Peter Ustinov, the great violinist Lord Yehudi Menuhin, the renowned actress Liv Ullman, and President Havel himself. We intended to make this a club of independent thinkers and activists who can think as well as feel the way to the better way to think and act. They did not adopt the everyday rationality, they had a deeper, wiser mindset.

A few months after my visit to India, we organized the first meeting of the Club of Budapest with its first members. The Dalai Lama himself attended as an Honorary Member. (We called our members "Honorary Members" to honor their commitment to the Club's objectives.) We reviewed the draft of what we called "Manifesto on the Spirit of Planetary Consciousness" and it was adopted by the Club of Budapest.

David
This brings us to the question of wisdom. Do you agree that the greater our material power, the greater the need for wisdom? Today, this need would be greater than it has ever been.

Ervin

The need for wisdom in today's world is because we need to stop behaving insanely. There is a tremendous amount of insane behavior in the world. Everybody is rushing. People rushing for power and profit and adopting a very self-centred attitude.

This narrowly chauvinistic attitude was typified by the much repeated motto of Donald Trump. "America first." He didn't say who comes second, and even if there is a second. Nobody comes after us, it's us all the way. You can't have a complex and highly interacting international system operating on this principle. This would only suboptimize the system and create conflict and inequality.

We need a new wisdom, which is also a very old wisdom. To adopt it we need to go back, but not back to where we were, but back to what we were meant to be, part of the evolving system of life on the planet. We need to aim for a more natural, a wiser way of thinking and being. A new normal, aligned with the norms of healthy life on the planet. This must a democratic, distributed power normal, a sanely self-governing normal. This is very much missing in the contemporary world. Governance is still hierarchical and self-centered. There is a major wisdom gap in the human world.

David

The way that you put this in your book is a coherence lag—a lag in coherence within and among people. We are taking too much time to become sufficiently coherent. Maybe you could say a little more about coherence and how you see our approach to it.

Ervin

Very well, let me go back to our mutual friend and mentor, the Nobel physicist Ilya Prigogine. He showed that a complex system can function and survive only if all its elements are working together. This means that the all its elements are responsive to one another and together work to maintain the system in the intrinsically unstable condition in which all living systems subsist, namely, far from thermodynamic equilibrium.

Coherence means that all elements of the system are responsive to each other. Together the parts make a whole that is more than the sum of its parts. Coherence is absolutely a key feature of evolving systems. The drive toward coherence appears to be inherent in the universe.

This systemic cooperation is the gist of coherence. Einstein said that the most remarkable thing about the universe is that it's so coherent that we can understand it. The coherence of the living world is the most striking feature of this region of the universe.

The drive or impetus toward coherence is there already at the most basic level of the physical world, the level of atoms. The so-called Pauli Exclusion Principle shows that atoms are not created simply by adding electrons to nucleons, but by integrating electrons and nucleons in a coherent structure. This is how stable "neutral" atoms emerge in the particle soup of the early universe, and how they further interact to create multi-atomic molecular structures.

We can also describe the process by saying that coherence emerges as laws of nature limit the degrees of freedom by which particles relate to each other and the rest of the universe. Coherence is a limitation, but it is the limitation of chaos and not of order. It is the emergence of order. Creating coherent systems is the outcome of long-term cosmological processes, and it may be the goal of evolution in the universe.

David

This vision was very much shared by Willis Harman, the founding president of the Institute of Noetic Sciences. He was thinking absolutely along these lines. He was one of the few people who had the vision to bring together a group of leading process thinkers. That group was a very great influence in my life.

Now, Ervin, there is an important component which I know is of interest to everybody, which is how you distinguish between a biophysical and a psychophysical universe in reference to the presence of consciousness.

Ervin

Yes, David, this is a tremendously important question. It is fundamental for understanding who we are, entities emerging and evolving in the universe. I believe we can best describe the true nature of the universe as "psychophysical." The universe is both physical and psychological—spiritual. It has aspects that appear to us as spiritual, mind-like, and aspects that appear to us as physical, matter-like. But these are aspects, appearances, not basic realities. Referring to the universe as physical does not mean that the universe would be material. Max Planck in one of his last talks in Florence, Italy, said that after forty years of studying the most basic elements of the universe, namely atoms, he can say that there is no such thing as matter. There are no hard, separate things that would move according to mechanistic laws. The universe is wider and deeper than that.

With the advent of quantum physics, we can now say what the universe truly is with more confidence than ever before. We can now conceive of the universe as a field of vibration, of which the elements are individual vibrations that resonate together and create more complex clusters of vibration. No vibration and cluster of vibration is entirely separate from any other. There is nothing here that would be material, although relatively stable clusters of vibration may appear matter-like. Astronomer Sir James Jeans said it well: the universe is basically more like a big thought, than like a big rock.

David

Indeed. Planck added that mind is the matrix of all matter. We have some changes in our concepts awaiting us, and they are not easy. It may take a lifetime to comprehend and complete these changes.

Ervin

We now realize that there is something like a cosmic consciousness pervading the universe, a consciousness that is the template for all the things that manifest in space and time. What we perceive are elements, projections, manifestations of this cosmic consciousness.

They are projections or manifestations of what David Bohm called the explicate order. This order is not the ultimately reality; it is created by the dimension that is ultimately real—the implicate order. The implicate order is more like a cosmic consciousness than any material sphere of being.

David

Exactly. There is a resurgent interest in David Bohm's work. We hosted a film, *Infinite Potential*, which deals with it and it had half a million viewers.

Ervin

The emerging insight is that we are part of each, and part of a larger whole. Then there is nobody who could be said to be "other." Nobody is a "stranger" or "foreigner." We are all living systems, part of the web of life on earth.

If this is so, it is ridiculous to carry on like we have been, fighting each other and harming, often killing, each other. Contemporary politics is largely a fight for power. The way politicians fight, it's not so much *for* something, but *against* something—against competitors, against anybody who seeks to be richer or more powerful.

David

These are very outdated concepts which need to disappear as we move into a better world.

To create a better world—this brings me to what you call timely wisdom. Reviewing, rethinking, challenging, expanding our sense of identity, and reconnecting to the source. Is there anything you'd like to highlight from the wide range of wisdom principles? Expanding our identity and reconnecting with the source could be the key.

Ervin

That's exactly that what they are. The urgent need is for each of us to expand his or her identity. Like everyone else, I am not just

Hungarian or American or Italian or whatever. I am also that, but first of all I am a human being and a human being is not categorically "other" than any other human being, and even any other form of life. We are all living beings on this planet, part of an evolving system and we evolve together. We need to expand our identity to encompass all living beings on earth.

David
Yes, but how do we do that? What are your thoughts and recommendations?

Ervin
To expand our identity we have to let go of the ideas that frame our past, narrow identity. We have to let go of many of the things we have been taught in school and were told at home and in our families. They are no longer valid sources and resources for living in this interacting and interdependent social and ecological system.

The ideas we have been living by have become outdated. And some of them are irrational. I have listed some of these irrational ideas and the behaviors they inspire in my book on the Wisdom Principles. The timely wisdom is asking, does what we think and do really make sense?

I have started my mature life by questioning all the ideas that govern our thinking and behaving. It was not enough for me to be just playing the piano and doing my best to please audiences. I wanted to find out what is the sense of it all. What is the purpose of my life?

Questioning is important, because if you find what is wrong, you open the door to asking, what can I do about it? Asking this question and taking it seriously is the beginning of wisdom.

David
Indeed, and do you still propose what might be called a philosophy of progress? You probably would prefer to use the term evolution, rather than progress?

Ervin

I think real progress is to go with evolution. As young people would say, it is to go with the force. The force of evolution; a subtle but real impetus. This impetus is in us. It is a tropism or attraction toward wholeness and coherence. I use the term originally suggested by Stanislav Grof—*holotropism*. This wholeness-oriented impetus is an inner drive towards coherence, manifesting as a feeling of love for all things—universal love.

The holotropism I am speaking of is encoded in our cells, in our hearts and brains. We need to recognize it and bring it to the level where it can influence our thinking and our actions.

David

I want to ask you whether you think we should come together and create a kind of wisdom council.

Ervin

Absolutely, yes. There are already examples of such councils, for example, the Elders convened by Nelson Mandela and Jane Goodall, among others. I was heading for many years the Wisdom Council of The Club of Budapest. We had many important meetings. Presently I am involved in the creation of the Wisdom Council of the new internet platform "PeopleTogether." These initiatives are important because we are in desperate need of the kind of guidance that could be provided by the members of such councils.

David

E. F. Schumacher said that humanity is now too clever to survive without wisdom. We hope that humanity can come up with the wisdom to survive.

Let me come back to something particularly important in what you said earlier. It is that it's not a question of coming up with a rational plan and imposing it from above, but rather of tuning in to what is emerging and empowering that.

Ervin

This is the gist of genuine leadership. Leadership means picking up what is moving people forward and empowering it, guided by the universal impetus for evolution, "the force."

We are part of the force. I know, that sounds poetic and soft, but this impetus is working in us and for us. It is more effective when we know it and can follow it.

David

Be healthy and enable others to be healthy as well is the task before us. That's the task. To me, it is the purpose of why we're here. I'm sure all deep-thinking people agree with that.

In the late nineteenth century, there was a New Thought movement and one of the metaphors that it used in particular came from Thomas Troward, a circuit judge in India. He said that we all have the experience of being a center. This is to say "I am"—to be aware of oneself. What this says to me is that there's only one center, but we are all cells of the microcosms within the same center. Does this resonate with you?

Ervin

Quantum scientists are telling us that we are really one. And many consciousness researchers say that the one of which we are a part is a cosmic consciousness. Here I introduce a daring idea that I am now convinced of. *The cosmos is consciousness.* This is the ultimate insight. We are elements, projections, manifestations of the cosmic consciousness. With this insight, the pernicious and mistaken fragmentation of the world into separate and often opposing people, organizations, and states can be overcome.

David

There are obstructions in the progress of the evolution of which we are a part. Our evolution is a very nonlinear process. Is the introduction of dysfunction a necessary condition for achieving coherence?

Ervin

There are always disruptions and disruption, that is part of the process. The course of evolution is not predetermined. Disruption, temporary obstruction brings about change, reinforces change, opens up the system to novelty.

David

What is your advice to the younger people who are listening to us or reading these lines? What advice would you give to parents and grandparents to pass on to the next generation?

Ervin

Open up. Take the natural, evolutionary road ahead. It is the road to a miraculous world. A world that merits being loved and being promoted to the next stage of its evolution. Open up to it. Receive it. Open your heart to receive it even before it is received and articulated by your brain. Become what you can and now need to become. Join with others to create a coherent world, a larger whole of which you can be a positive part. Strive for power, but not for power over others, but the power to be...so others could also be. Strive to thrive together, and not just to barely survive.

My advice is to become what you were meant to be. Become a person who loves other people, nature, and the universe, and loves them unconditionally. One who knows and feels that he or she is part of this evolving, miraculously coherent universe. One who strives to make humankind thrive by spreading human kindness. This is a meaningful goal for all of us.

David

A great advice for all of us. Thank you so much, Ervin, for speaking to us about your life, your thinking, and your hopes for a better world.

Made in the USA
Monee, IL
24 April 2022

95338388R00083